NEW PRINCIPLES OF ORIGINS AND EVOLUTION:

Revolutionary Paradigms of Beauty, Power and Precision

Alexander A. Scarborough

Featuring

The New Fourth Law of Planetary Motion:
How the planets attained their orbital spacing around our Sun.

© 2001, 2002 by Alexander A. Scarborough. All rights reserved.

No part of this book may be reproduced, stored in a retrieval system, or transmitted by any means, electronic, mechanical, photocopying, recording, or otherwise, without written permission from the author.

ISBN: 0-7596-8417-0

This book is printed on acid free paper.

First published in 2001 by Ander Publications, 202 View Pointe Lane, LaGrange, Georgia 30241

ENERGY SERIES

First Edition: Fuels: A New Theory (1975)
Second Edition: Fuels: A New Theory (1975)
Third Edition: Undermining the Energy Crisis (1977)
Fourth Edition: Undermining the Energy Crisis (1979)
Fifth Edition: From Void to Energy to Universe (1980)
(Manuscript unpublished)
Sixth Edition: New Concepts of Origins: With White Fire Laden (1986)
Seventh Edition: The I-T-E-M Connection: How Planet Earth and Its Systems Were Made by Means of Natural Laws (1991)
Eighth Edition: The Spacing of Planets: The Solution to a 400-Year Mystery (1996)
Ninth Edition: New Principles of Origins & Evolution: Revolutionary Paradigms of Beauty, Power and Precision (2001)

CONTENTS

Foreword vii
Author's Philosophy of Science and Religion ix
Preface xi

Chapter I. On the Spacing Of Planets: The New Fourth Law of Planetary Motion 1

Abstract. Introduction. Introduction to the Geometric Solution to the Fourth Law of Planetary Motion. Extrasolar Systems. Introduction to the Five Stages of Planetary Evolution. Introduction to the FLINE Paradigm of Planetary Origins and Evolution. Summary. More on the Spacing of Planets: History's Firm Foundation. The New Fourth Law of Planetary Motion. More on Geometric Origins. How Stable is the SS? Do Other SSs Exist? The Ages of Stars and Galaxies. Two More Extrasolar Planets. Extrasolar Systems: How and Why They Differ From Our SS.
Letters and Memos: Migrating Planets; Response. Images of Suspected Planet. First Multiple Planet System. Recent Corroborative Evidence. Plentiful Planets. Shaking Up a Nursery of Giant Planets. A Field Guide to the New Planets. Journal of the Royal Astronomical Society.

Chapter II. How Planets Evolve 33

The FLINE Paradigm. Abstract. Introduction to Planetary Evolution. Dinosaurs: The Reason for Their Extinction. Update on Dinosaurs' Extinction. Four Original Clues to Nuclear Cores. Dinosaurs: Inside and Out and Then There Were None. Recent Clues to Nuclear Cores in Planetary Spheres. More Evidence Favoring Earth's Energy Core. Corroborative Evidence for Nucleosynthesis. Earth Story - A TV Series. Ringing Earth's Bell. Fusion Energy and Magnetism. Why Electromagnetic Field Strengths Vary. The Mysteries of Earthquakes. The Warning of Precursory Signals. The Kobe Earthquake Signals. Depth as a Safety Factor. Why the Explosions? The Tiny Mystery of Polonium Haloes: Creationism, Big Bang or the FLINE Model? Galileo's Stunning Probe Into Jupiter. The FLINE Paradigm: Creation by Means of Natural Laws; Abstract. Evolutionary Background of the FLINE Model. The Five Stages of Planetary Evolution. The Five Fundamental Principles of Origins and Evolution of Planetary Systems. Origins of SSs and the Evolution of Planets: FLINE vs. Accretion. Accretion or Natural Laws? Conclusion.
Letters and Memos: Natural Nuclear Reactors. PHYSICS TODAY. AGU Membership. UGA: A Field Guide to the New Planets. Royal Astronomical Society. Methane Hydrates. Turning Stars Into Gold. Astronomers Detect Birth of Planets. Infrared Gleam Stamps Brown Dwarfs as Stars.

Chapter III. How Earth's Systems Evolved 71

The Myth of Fossil Fuels. How Hydrocarbon Fuels Formed in Earth's Crust. The Three-Layer Systems in Earth's Crust. How Coal Formed From Petroleum. How Petroleum Formed From Gas. Source of the Methane Gas. Evidence for the Making of Abiogenic Methane. Rise and Decline of the Abiogenic Fuels Theory. Revival of the Abiogenic Fuels Theory. From Energy to Matter to Life. The Universal Law of Creation of Matter. CONNECTIONS: Gaia, Natural Selection and the LB/FLINE Model.
Letters and Memos: The World Has More Oil, Not Less. Resources and Ramifications. Why Earth Will Never Run Out of Oil. It's No Crude Joke: This Oil Field Grows Even as it's Tapped. Think Ahead. Energy Fuels Background. News Release. Why We'll Never Run Out of Oil. Physics News. Star Material Discovered in South Pacific. Traylor: Three Questions. Powering the Next Century. A Secure Conclusion.

Chapter IV. Moons, Planetary Rings, Comets And Asteroids 95
Origin of Our Moon. Why We Have a Moon. Recent Developments Concerning the Lunar Cataclysm Hypothesis. New Evidence Substantiating the FLINE Model of Creation of the Rings and Moons of Jupiter. The Far Side of the Moon. How Planetary Rings Were Formed. Jupiter's Rings: A Giant Leap Forward. Comet Halley and Its Last Farewell. Comets and Asteroids: Keys to Planetary Origins? Hale-Bopp: A Great Comet. How Comets Stay Active in Near Absolute Zero Cold. Comet Collision With Jupiter: Some Stunning Results. Royal Astronomical Society. Response. Belief vs. Science: Which is More Important?

Letters and Memos: Lunar Prospector Mapping. Comet/Moon Sampling. Was Chicken Little Right? Stardust Spacecraft: Comets. Jupiter's Moon Io. Comments on Comets. More on Comets. Lunar Prospector Findings on the Moon's Metal Core. Movements of Jupiter? The Spacing of Planets. Comet Hale-Bopp, Ganymede, Io, and Buckyballs. A Question of Ethics in Science. Addendum.

Chapter V. A Big Bang Or Little Bangs? 127
The Three Key Observations. Questions About an Expanding Universe. The Cosmic Microwave Background: The 2.7 K Radiation. A Closer Look at the CMB Radiation. The Relative Abundance of Elements via the BB. The Fiery Little Bangs Theory: A Plausible Alternative. More on Galaxies. How Stars Form. Cosmic Misfits. The Irony of Iron in Stars and Planets. Redshifts: A Shaky Measuring Rod for Astronomical Distances? Should We Believe the BB Scenario? Solutions to Anomalies: A Summary List. Black Hole in Milky Way Center.

Letters and Memos: A New Look at Black Holes. Exploding Stars Point to a Universal Repulsive Force. Water Found in Orion. The Shoulders of Giants. More on Black Holes. Electric Space. Common Sense Science. Cosmic Motion Revealed. The Recycling Universe. A Different Approach to Cosmology. The Heart of the Matter. Making the Stuff of the Big Bang. Not to Teach Theories as Facts. Origins of Solar Systems and the Evolution of Planets: The Beauty and the Precision. A Plea to the AAAS. The Spacing of Planets. The Universe in a Sphere. On the Spacing of Planets. Space Observatory Shows Black Holes Once Dominated. An Opportunities-Based Science Budget. Another Plea to the AAAS. The Genesis Mission to the Sun. A Time for Change. About the Author.

FOREWORD

During the past quarter century (1975-2000), a revolutionary concept of dynamic origins and evolution has been painstakingly put together. Like pieces of a jigsaw puzzle, a multitude of facts interlock precisely to reveal the sheer beauty, power and precision of definitive and testable, but unorthodox, concepts of universal origins and evolution: the Little Bangs (LB) and the FLINE model of origins of solar systems and evolution of planets, moons, stars, etc.

An important lesson to be learned from the new LB/FLINE model is that any discovery in the physical sciences should be interpreted in more than one perspective before judging which interpretation has truer scientific validity. Examples of this lesson spring from the discoveries of Copernicus, Galileo, Descartes, Dutton, Darwin *et al*. — men whose discoveries have outlived the popular beliefs of their time.

Upton Sinclair once wrote, "It is difficult to get a man to understand something when his salary depends on his not understanding it." Unfortunately, in the strict peer review system of science, new ideas in opposition to beliefs of reviewers schooled in current dogma are rejected out-of-hand in spite of sufficient substantiated evidence and the absence of speculation. Thus, prevailing beliefs are shielded against the intrusion of new ideas that indicate the need for a change in the direction of scientific thought; scientists are permitted little, if any, opportunity to challenge the scientific validity of new ideas or to benefit from them. The damages wrought by such policy, as history teaches, can be huge.

To quote World magazine: "As Thomas Kuhn has shown in his book, *The Structure of Scientific Revolutions*, when scientists discover more data, they must discard old models, as new models take shape. But unfortunately, between the models comes virulent controversy. Whenever the old model becomes exhausted, Kuhn showed, the scientific establishment acts in a predictable way. First, it stretches the old model beyond plausibility in an attempt to account for the new data. Then the scientific community lashes out at those who dare challenge the existing model" — a procedure aptly named the Galileo treatment.

Today, new data accumulated during the past quarter-century poses a serious challenge to the Big Bang/Accertion model of universal origins and evolution. Forced to adhere strictly to the prevailing model, scientists must struggle with, and often strangle on, speculative interpretations of important discoveries and critical data that always fit more logically into the LB/FLINE model. But scientists usually cannot see the evidence differently until they change their interpretive framework. Then new concepts in this revolutionary publication provide a convincing basis for that change.

Throughout the Universe, nothing escapes Nature's strict laws that make it possible for mankind eventually to understand how everything came into being. Essential to that understanding is sufficient knowledge to eliminate the need for speculation that too often leads to misinterpretations of important findings. As history teaches, new ideas, generally known as breakthroughs, precede new knowledge, and knowledge begets knowledge. Adhering to the precise laws of Nature and soundly backed by substantiated evidence, the revolutionary FLINE model is the only known concept of planetary origins, orbital spacing and evolution that seems capable of surviving the test of time.

The struggles involved in researching and preserving these new findings in the face of strongly entrenched beliefs and under adverse conditions therefrom seem worthy of recording for posterity. To give the reader a broader historical and insightful picture of the slowly developing situation in which discouraging rejections and expensive struggles were common, the contents of some typical correspondence, usually concerning new discoveries, is presented in its original letters form in each chapter. But most of the newer discoveries are blended precisely into the cumulative text of the past quarter century (1975-2001). Advocates of the definitive FLINE model foresee the time when it will have an impact on scientific beliefs equal to that of the Copernican idea of our heliocentric SS described in his masterpiece, *Concerning the Revolutions of the Celestial Spheres* (1543). If so, building onto these new findings during the 21st century should bring the exciting knowledge of our origins (i.e., the Copernican Revolution) full circle.

AUTHOR'S PHILOSOPHY OF SCIENCE AND RELIGION

Einstein's famous 1905 formula $E = mc^2$ reveals that the Universe is composed only of two basic, and totally interchangeable, substances: energy and matter. An ever-fuller appreciation of this most fundamental principle of the Universe will enable scientists to understand the intimate relationships of the precisely interwoven laws of chemistry, physics and mathematics that eventually will reveal how the Universe functions.

God established these inviolate laws and then instilled in mankind the brain-power to discover them one by one — a process we call science. Mankind's souls were instilled to discover and enact the purposes of our existence in the Universe —a process we call religion.

PREFACE

The most fundamental mystery in science and in human imagination is the nature of our origins. Throughout civilization, humankind has been obsessed with trying to understand how and why everything came into being. The concepts in this book build on the solid foundation of understanding established by many great names from the past. Each of their discoveries dispelled established myths pertaining to our origins.

This revolutionary work takes us from the Copernican idea of our Sun-centered Solar System (SS) to the recent discoveries of giant planets in other distant solar systems as it seeks do no less than rebuild the foundation of knowledge of the origins and evolution of planets. It offers new answers to questions that have puzzled philosophers for centuries: How did Planet Earth come into being? How and why did the planetary orbits of our Solar System form in a mathematical pattern? Why did this enigmatic solution to the spacing of our planets remain a mystery for 400 years after it first eluded Johannes Kepler in 1595? Will this solution cause scientists to rethink their beliefs about the origins and evolution of planets?

These profound questions are addressed with powerful substantiated evidence that voids the need for speculative uncertainties now common in current theories of planet formation. The book aims to advance scholarship and to enlighten readers with no background in science and philosophy. The first chapter opens with an insightful and startling account of the enigmatic solution to the new Fourth Law of Planetary Motion explaining how the nebulous planetary masses attained their orbital spacing around the Sun. In conjunction with Kepler's Three Laws of Planetary Motion, the Four Laws (FL) reveal the explosive, dynamic origin of our Solar System some five billion years ago, and thus challenge the modified Laplace accretion concept of planet formation.

The book gives definitive insights into how and why each planet evolves through five common stages of planetary evolution in full accord with Einstein's famous formula and all natural laws. This new FLINE model makes it easy for readers to understand the origin and evolution of planetary matter; e.g., hydrocarbon fuels (gas, petroleum, coal) are formed via IN and E - not from fossils - and thereby illustrate how Planet Earth and all other planets evolve via IN and E: Nature's two inseparable principles.

Major myths are challenged and decisive discoveries are brought full circle into the 21st century. The nature of comets, planetary rings and asteroids are explored in depth. The FLINE Paradigm is a profound work that provides what philosophers have sought through the ages: a definitive foundation for understanding the origins of solar systems and the evolution of planets via the new FLINE model. To understand anything and everything about planets, we must first understand their three governing and inseparable principles: The Four Laws of Planetary Motion (FL), Internal Nucleosynthesis (IN) and Evolution (E).

For the first time, the geometric, fiery origin of the SS, the spacing of planets and the subsequent evolution of all planets and planetary systems via internal nucleosynthesis (IN) can be explained within the scope (and by the means) of natural laws in a beautiful continuity of substantiated evidence. The FLINE model is the only concept capable of making such claims and backing them with facts.

The ideas in this book are the culmination of a quarter century of painstaking research into past and current beliefs about universal and planetary origins. Surprisingly, some little-known, but accurate interpretations and a few brilliant ideas emerge from among the relevant hypotheses suggested throughout the pages of scientific history. Sadly, some of the most astute observations and accurate interpretations have lain buried beneath the onslaught of more popular, but erroneous beliefs.

During the past quarter century of its development and presentations of evolving facets (and of the completed concept during the past five years) to scientific organizations, the definitive FLINE model has not encountered any valid argument against its basic principles - nor does that seem likely to happen.

The important thing is not to stop questioning.

Albert Einstein

…a factual, non-speculative easy-to-understand concept that makes a lot of sense. It provides answers to scientists who rethink current beliefs about planetary origins and evolution.

Dr. Morton Reed, Professor of Thermodynamics

Chapter I

ON THE SPACING OF PLANETS:

The New Fourth Law of Planetary Motion

Abstract

All efforts to solve the mystery of the orbital spacing of the planets around our Sun have failed to produce a definitive solution. To understand the origins of solar systems and the evolution of planets and life, the critical connection between the spacing of our planets and the forces that drive their evolution must be explained beyond doubt. Prevailing concepts fail to explain this inseparable connection. The basic key to this relationship resides in Kepler's First Three Laws of Planetary Motion and the solution to the new Fourth Law. Together, the Four Laws reveal the dynamic manner in which our Solar System (and all solar systems) came into existence, thereby providing a solid foundation for understanding the forces that drive the evolution of planets and life. The mathematical solution is detailed.

Introduction

In his book, *The Divine Proportion*, H. E. Huntley wrote, "It was suggested in the early days of the present century that the Greek letter Phi should be adopted to designate the golden ratio. The ubiquity of Phi in mathematics aroused the interest of many mathematicians in the Middle Ages and during the Renaissance. In 1509 there was published a dissertation by Luca Paciola, *De Divina Proportione*, which was illustrated by Leonardo de Vinci. Reproduced in a handsome edition in 1956, it is a 'fascinating compendium of Phi's appearance in various plain and solid figures.' We shall in following chapters come across many examples of the appearance of Phi in unexpected places." The examples are indeed numerous. (1)

C. Arthur Coan best described the Golden Ratio when he wrote: "Nature uses this as her most indispensable measuring rod, absolutely reliable, yet never without variety, producing perfect stability of purpose without the slightest risk of monotony... We shall find it flung broadcast throughout Nature." Some examples are found in the spacing of leaves on a stem, the logarithmic spiral or nautilus, and the designs of snowflakes. (1)

In the early 17th century, the brilliant astronomer, Johannes Kepler, published the First Three Laws of Planetary Motion:

1. The orbit of a planet is an ellipse, of which one focus is the Sun.
2. For each planet, a straight line joining it to the Sun sweeps over an equal area in equal time.
3. For each planet, the cube of its distance from the Sun is equal to the square of its time of revolution. (i.e., the greater the distance, the slower the velocity).

Some 19 years earlier, in 1595, to be exact, Kepler had recognized the geometric pattern in the spacing of the six then-known planets, and had attempted to explain this anomaly by placing a geometric configuration between each of the orbits. His model has the five geometric figures between the six orbits. The largest one is the cube between Jupiter and Saturn. Of course, this reveals nothing, except that it does acknowledge the fact that planetary orbits are geometrically spaced around the Sun. However, it does not explain why or how they became spaced in the geometric pattern (1).

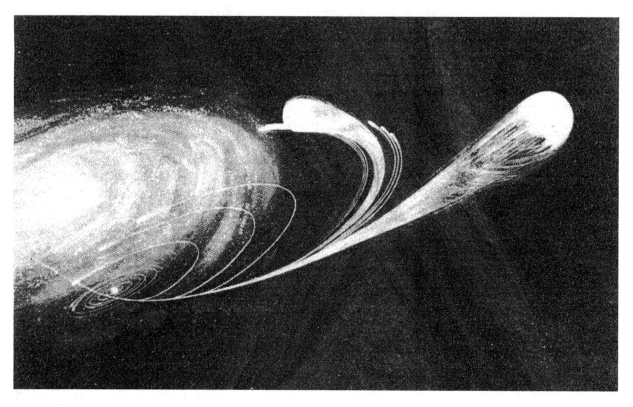

BIRTH OF THE SOLAR SYSTEM
This painting sketched by the author and painted by Charles Warner, shows the masses of Jupiter and its moons breaking off from the mother mass, which continues on its path to form Saturn, Uranus, Neptune and Pluto.
Figure 1

Ironically, Kepler wrote, "Geometry has two great treasures: one is the theorem of Pythagoras; the other the division of a line into extreme and mean ratio. The first we may compare to a measure of gold; the second we may name a precious jewel."

Kepler's precious jewel had been known to the ancient Greeks as the Golden Ratio in 500 B.C. They built their beautiful Parthenon in the GR proportion, the basis of the Phi geometry that pervades throughout Nature, and, ironically, is the vital key to solving the mystery of how and why the planets are spaced in a geometric pattern around our Sun.

In view of his precious jewel and all that was known to Kepler about the GR or Phi geometry, it is puzzling as to why he did not use it to solve the mystery of the geometric spacing of the planets. Even more puzzling is why this mystery persisted for 400 years after Kepler's initial attempts in 1595 to solve it.

To understand the seven graphs (1) that follow, it would help to visualize this picture (Fig. 1) in your mind. In it, two masses of fiery Sun-like energy, set in motion by a powerful explosion of a much larger energy mass (e.g.; quasar or black hole), are moving rapidly near the outer edge of the Milky Way galaxy. The faster, smaller mass passes the larger Sun mass shown in the lower left corner, and is diverted at a 14° angle as it passes by the larger mass. The forces of gravity act on the smaller mass to cause breakouts of smaller Sun-like masses that later will evolve into planets. As you visualize this scenario, keep in mind two critical things: (1) when it passes the larger Sun mass, the mother mass goes into a decelerating-velocity mode, due to the powerful gravity of the Sun, and (2) along the way, each nebulous planetary mass breaks away from the decelerating mother mass(es) at the precise GR points.

Introduction to the Geometric Solution to the Fourth Law of Planetary Motion

The following 3 sets of a total of 7 diagrams reveal how the 12 nebulous planetary masses were placed in a geometric pattern around our Sun in full accord with the GR. (1). They include the subsequent changes in each planet's orbital position and velocity since the fiery origin of our SS some five billion years ago. You will see that the three sets of geometric diagrams (of 1980-1995) corroborate each other in validation of this solution to the new Fourth Law of Planetary Motion that first eluded Kepler in 1595.

In the first diagram (SS-1) of the first set, the hypotenuse represents the path of the decelerating mother planetary mass as it moves along the incline while the Sun moves at a steady pace along the X-axis. The smooth curve on the right is its apparent path as viewed from the moving Sun during the fiery layout of the planetary orbits in their original GR positions. The warped curve shows their current positions. The two curves clearly reveal that the plot of current orbital positions is a warped version of the original GR positions. The deviations of the planets from their original GR positions, as tabulated here, are attributed primarily to gravity forces that have occurred over a period of some five billion years. The key formulae for the original and current positions are shown in the heading above the recorded data. Now let's expand the small distance between the Sun and Jupiter to show the inner planets in more detail.

In the expanded second diagram (SS-2) of the first set, the mother mass moves up the hypotenuse at a decelerating pace. However, when viewed from the moving Sun during the original layout of the orbits of the inner planetary masses of Mercury through Jupiter, the mother mass appeared to move along the path of the outer curve on the right — an optical illusion created by the relative motions of the two primary masses. The inner curve is derived from the current positions of the planets.

The first diagram (SS-3) of the second set shows how each inner planetary mass was pulled into its GR position by a combination of the gravities of the Sun and the preceding planetary mass acting on it at the tangent line T. The difference of 14° between Y and Y' is the angle of diversion of the mother mass as it passed the Sun and continued moving away at a decelerating pace (a critical factor).

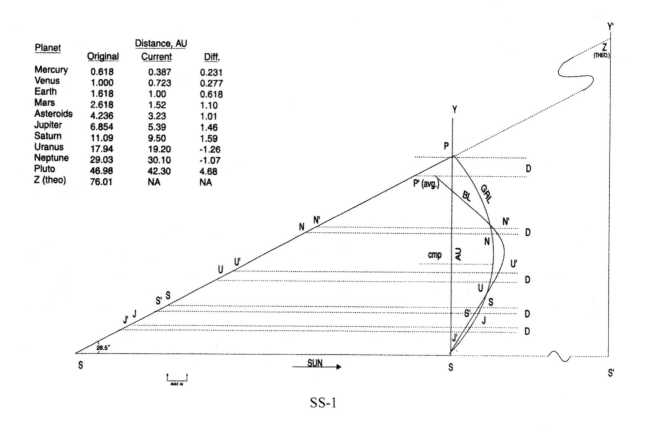

SS-1

New Principles of Origins and Evolution
Revolutionary Paradigms of Beauty, Power and Precision

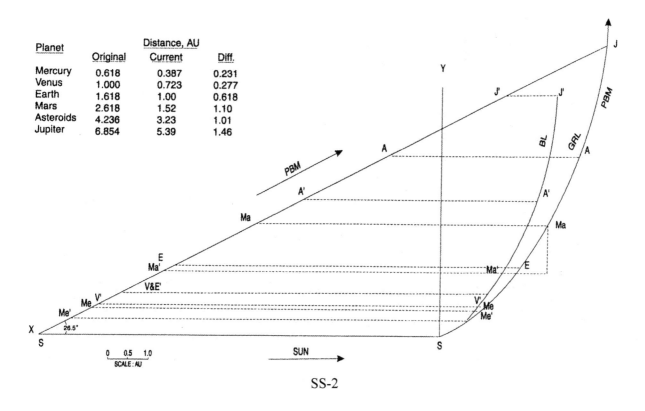

SS-2

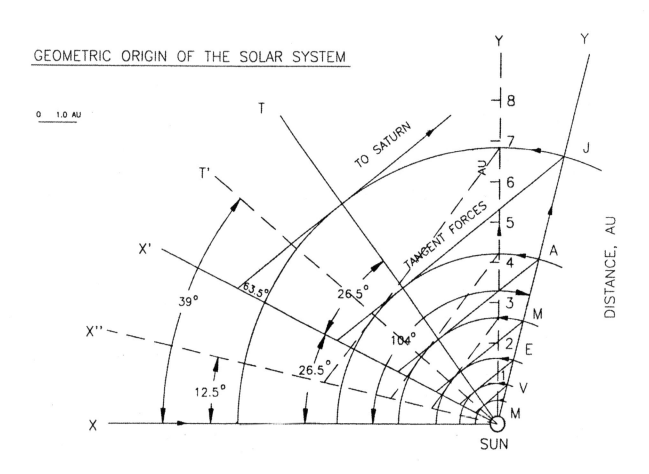

ILLUSTRATES HOW THE INNER PLANTS WERE PLACED IN ORBITS VIA THE GOLDEN RATIO GEOMETRY OF THE FOURTH LAW OF PLANETARY MOON.

SS-3

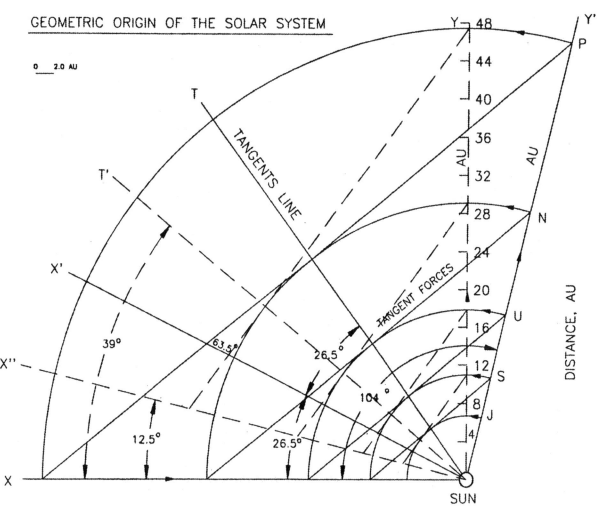

ILLUSTRATES HOW THE OUTER PLANTS WERE PLACED IN ORBITS VIA THE GOLDEN RATIO GEOMETRY OF THE FOURTH LAW OF PLANETARY MOTION.

SS-4

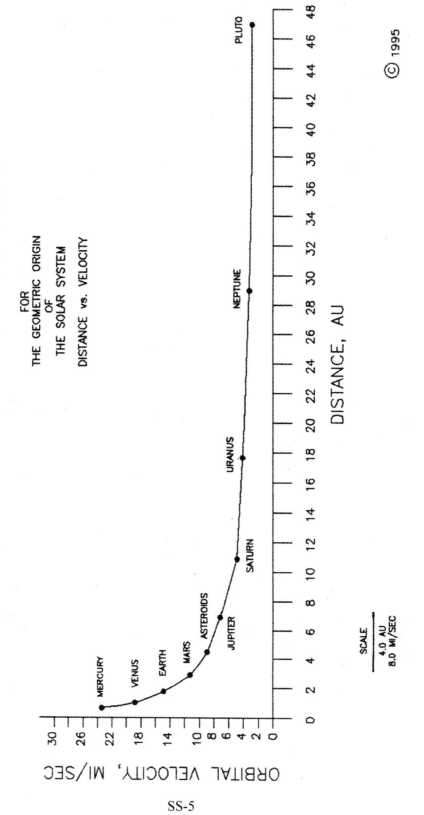

New Principles of Origins and Evolution
Revolutionary Paradigms of Beauty, Power and Precision

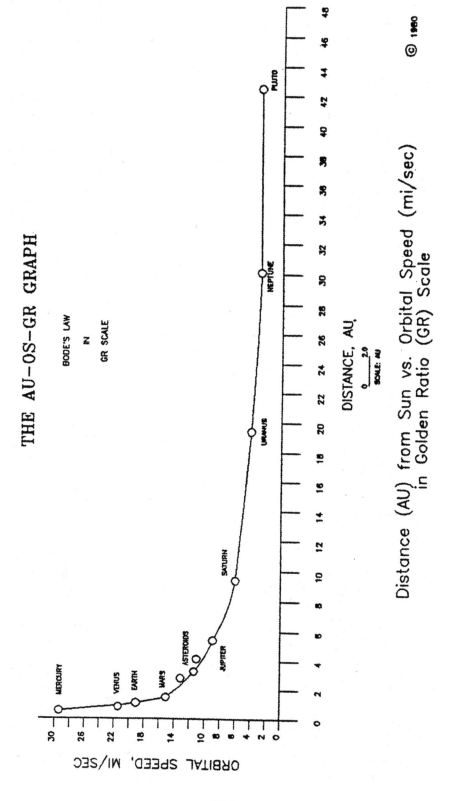

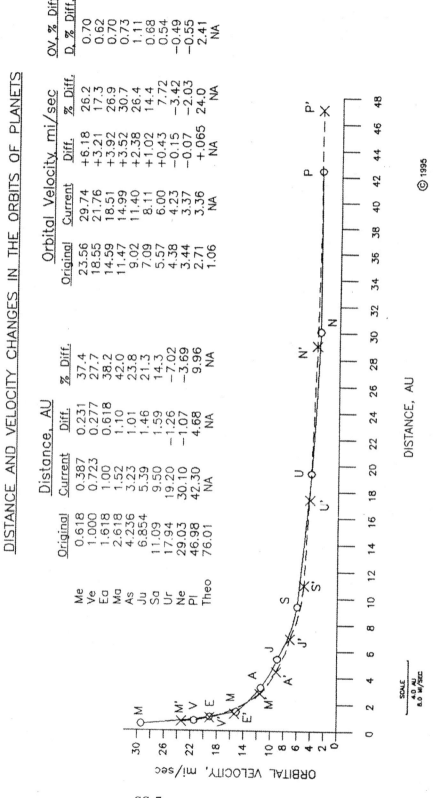

The second diagram (SS-4) of the second set, drawn on a smaller scale to include all the planets, illustrates how the outer planets were pulled into orbits in the same manner as was shown in the previous graph of the inner planets. Note that the distance of any exit point is always 1.618034 (GR) times the distance of the previous exit point. These GR distances are maintained from Mercury to Pluto. And moving back toward the Sun in the opposite direction, we see that the distance between any two spheres is always 0.618034 (GR) times the previous distance between two spheres.

In the first diagram (SS-5) of the third set, the current positions of the planets are plotted against their orbital velocities. In this third set consisting of three graphs, the distances in AU of the planets from the Sun are plotted along the X-axis, and the orbital velocities are shown on the Y-axis.

In the second diagram (SS-6) of the third set, the original GR, or Phi, positions of the planets are plotted against their original orbital velocities, which were calculated by Kepler's Third Law of Planetary Motion, $D^3 = T^2$.

In the third diagram (SS-7) of the third set, the two previous graphs are superimposed. They clearly reveal the differences between the original and current positions and between the original and current velocities of every planet. These differences are tabulated for easy references and comparisons. For example, note that Earth has moved 38.2% closer to the Sun while gaining 26.9% velocity. Uranus and Neptune have moved slightly away from the Sun by 7.02% and 3.69%, respectively, with decreases in velocities of 3.42% and 2.03%, respectively. Note that when a planet drifts toward the Sun, it must gain speed to stay in orbit; when it drifts away from the Sun, its velocity decreases, all in compliance with Kepler's Three Laws of Planetary Motion.

These results reveal that for any planet, its orbital position around the Sun at any time is a function of the distance-velocity-mass-time relationship to its central star.

From these seven graphs, the new Fourth Law of Planetary Motion that first eluded Kepler in 1595 can be stated: "The gravity-induced displacement of any planetary orbit since the geometric origin of the Solar System is determined by the difference between its current (BL) position and its original Phi (GR) position." The original velocity of each planet, of course, is determined by the Third Law of Planetary Motion, $D^3 = T^2$.

Extrasolar Systems

The Four Laws of Planetary Motion reveal the absolute necessity of a great momentum of smaller, faster mass interacting with a larger mass as it sped past our Sun to create the elliptical planetary orbits with geometric clockwork precision. Further, the solution to the Fourth Law reveals that in order to create a multiple-planets solar system like ours, in which the smaller planetary masses break from the speeding mother mass at the GR points along the way, the two initial masses must meet precise prerequisites of relative velocities, relative masses, and a specific distance apart to thereby effect a critical angle of diversion and subsequent breakup around a central star. Since it is a rarity for two masses to meet these exact prerequisites, the vast majority of solar systems will contain only one intact giant planetary mass circling its central star. (See Figure 5) At this time, astronomers have discovered about 50 of these exoplanetary solar systems, all at very great distances from our SS, and each with a planetary mass that failed to meet the prerequisites specified in the solution to the Fourth Law of Planetary Motion. As astronomers have acknowledged, each of these giant gaseous planets is much too close to its central star to have formed there via the prevailing accretion concept of planetary origins!

All solar systems comply with Kepler's First Three Laws. Whether the solution to the Fourth Law applies solely to our Solar System remains to be seen. But as we shall see, the sizes and gaseous nature of all the giant Jupiter-like planets, as well as the sizes and nature of all smaller planets and moons in our Solar System, can be explained by a new concept of planetary evolution that is based on a solid foundation: the Four Laws of Planetary Motion.

The Four Laws of Planetary Motion (FL) give vital clues to the dynamic origins of our solar system. But once in their GR orbits, how did the fiery Sun-like masses evolve into the planets we see today?

Descartes, in his *Principles of Philosophy*, defined Earth's interior as being Sun-like (Fig. 2). Later, Buffon, author of the 36-volume *Natural History*, speculated that fragments of the Sun formed into spheres that became planets. Considering the limited knowledge of those times, these two beliefs are amazing in how close they came to an elementary understanding of our planetary origins long before the fiery relationship between nuclear energy and atoms, as shown by Einstein's formula, $E = mc^2$, was discovered, and the processes of nucleosynthesis and

NUCLEAR FURNACE

Earth's nuclear core reactions may someday be duplicated by scientists to produce energy and desirable elements.

Figure 2

polymerization were recognized. Einstein's formula and the atomic bomb make it obvious that matter is simply another form of universal energy, and that all atomic matter was, and is, formed from energy. Such transformations of energy into matter, known as nucleosynthesis, must occur, of necessity, under extreme conditions of high temperatures and pressures within masses of energy. Various and varying combinations of these severe conditions determine the types and quantities of atomic elements produced via internal nucleosynthesis (IN) in any specific time frame by stars, planets, moons, and our Sun.

Elements ranging from hydrogen to iron have been detected in our Sun; no elements heavier than iron can be found therein. But heavier elements are produced in other stars and in planets. The more severe conditions created in Earth's thickly encapsulated core result in the production of all the elements, including the heaviest one, uranium.

The evolution (E) of planets (Fig. 3), in full accord with all natural laws, occurs at rates proportional to size: the smaller the planetary mass, the more rapid its transition through each of five common stages of evolution. Astute observations of the planets reveal excellent examples of the five ongoing stages of evolution: e.g., Energy (the Sun), Gaseous (Jupiter, Saturn), Transitional (Uranus, Neptune), Rocky (Earth, Venus), Inactive (or nearly so: Mercury, Mars, Pluto, Moon).

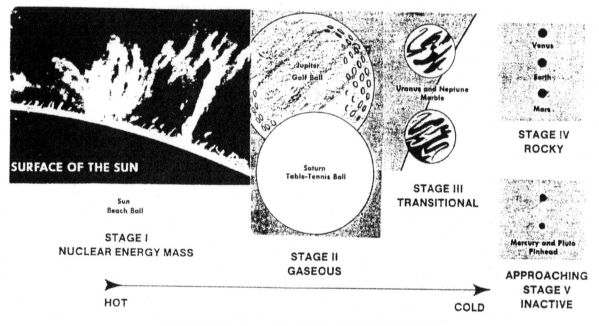

FIVE STAGES OF PLANETARY EVOLUTION
Figure 3

Note that the planets in the picture slowly evolve from hot-to-cold stages in full accord with size and in full compliance with the laws of thermodynamics. The transition from the first stage of evolution into the second, or gaseous, stage begins with the creation of vast amounts of the lightest and easiest-to-make atomic elements, primarily hydrogen and helium, the main elements found in the Sun. Countless numbers of atoms, ions and simple compounds eventually form a thick enveloping spherical blanket high above their source of creation. Simultaneously, as with our Sun, small percentages of atomic matter accumulate on the surface of the hot core. During eons of time, virgin matter billows from the manufacturing core, slowly filling the huge space between the core and the spherical blanket, thereby forming clouds of chemical vapors throughout the towering atmosphere. Scientists see these results in Jupiter and Saturn, and classify them as gaseous planets. Additionally, all known extrasolar planets, because of their huge sizes, are currently in this second stage of planetary evolution.

In the next phase of transition, the voluminous clouds of atmospheric matter gradually close in on the hot surface, and the precipitation of its chemical vapors is initiated. The precipitates are instantly repelled by the heat; the gigantic battles between the evaporation and condensation forces rage on for eons.

Trapped between the heat of the core and the super coldness of space, the virgin matter finally gains a toehold in the form of the first tenuous liquid layers, eventually followed by thin, tenuous crustal formations. Uranus and Neptune should be in this third, or transitional, stage of evolution.

During the fourth stage of planetary evolution, shallow seas cover the entire surface, cooling it sufficiently to permit formation of islands of more permanent crust, drifting on hot magma. The crust thickens as a function of time — an ongoing process for billions of years. Newly created matter from volcanic outpourings of lava, water, gases and compounds of great variety continually build and modify the atmospheres and land systems. The rocky crust grows ever thicker, while the atmosphere grows ever thinner. Earth is a prime example of this fourth, or rocky, stage of planetary evolution.

Planetary energy cannot last forever. When the hot energy core finally is depleted, the planet or moon in question becomes inactive — a dead sphere that cannot be revived. Mercury, Mars, our Moon and Pluto are examples of spheres that are in, or are entering, this fifth and final stage of planetary evolution.

Introduction to the FLINE Paradigm of Planetary Origins and Evolution.

With this information, we can put together the FLINE paradigm of planetary origins and evolution consisting of three chronological, inseparable and ongoing realities:

1. The Four Laws of Planetary Motion (FL)
2. Internal Nucleosynthesis (IN)
3. Evolution (E).

These three basic principles of the FLINE paradigm apply to all solar systems, except that single-planet systems disregard a part of the Fourth Law. No solar system like ours is possible without undergoing these inseparable realities. In any such system, the three realities remain interactive until, one by one, they reach their ending. They reveal that planetary evolution in any solar system is not possible with any type of core other than that hypothesized by Descartes as "Sun-like" — a concept now corroborated by abundant substantiated evidence. Such evidence mounts with each relevant discovery of planetary anomalies, both of Earth and of space probes. From energy mass to inactive spheres, every planet is a self-sustaining entity that creates its own atomic compositional matter via internal nucleosynthesis throughout the first four stages of its inevitable evolution. The essential relationship of internal nucleosynthesis to the evolution of stars, planets, moons, etc. (in which one cannot exist without the other) (1) poses a serious challenge to the idea that all light (and some heavy) atoms were created solely in a Big Bang. (2).

This FLINE model has an impeccable record of accurate predictions. Example: the discoveries in the 1980s of vast quantities of methane hydrates in Earth's crust had been predicted and explained in the 1970s by the Energy Fuels concept. Further, the FLINE model accurately predicted that the crashes of Comet SL9 fragments on Jupiter in 1994 would produce the most powerful explosions ever witnessed in our Solar System. The competing super computer predictions proved to be puny in comparison with the actual explosions. For example, Fragment G (Fig 4), exploding upon contact with the thin cloud-tops of Jupiter, released at least 6 million megatons of energy. (One megaton is the equivalent of one million tons of TNT.) Each of the 21 crashes were far more powerful nuclear explosions than the small one that hit Tunguska in 1908. However, both situations reveal the true nature of fiery comets: nuclear energy. Substantive evidence in the scientific literature strongly supports this conclusion.

The dramatic and costly explorations on Mars have revealed nothing that had not been predicted and explained by the FLINE model. The same statement holds true for the explorations of other planets and moons of our Solar System and for the findings about the 50 extrasolar systems.

The rapid progress being made in understanding the Sun's plasma is encouraging. (Ref: SN, 3/27/99, p.200). The results, when eventually applied to planetary cores, will corroborate Descartes' belief in Earth's "Sun-like" core that explains the five stages of planetary evolution, thus bringing this revolutionary concept full circle.

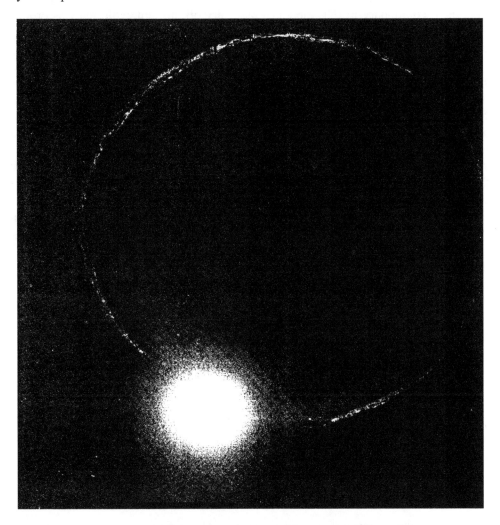

COMET S-L COLLISION WITH JUPITER.

A stunningly spectacular explosion of one fragment of Comet S-L 9 on Jupiter in July 1994 that far exceeded all predictions. Fragment G alone propelled a fireball thousands of kilometers above Jupiter's stratosphere and is thought to have yielded at least 6 million megatons of energy. A megaton is the equivalent of a million tons of TNT. The fragments exploded at the cloud tops, another crucial clue to their true nature: nuclear energy fireballs. The picture shows the actual explosion of a nuclear fireball rather than of a dirty rocky snowball. This information is devastating to the Big Bang theory.

Figure 4

Summary

Embracing the great ideas of Copernicus, Galileo, Kepler, Newton, Descartes, Buffon, Dutton, Einstein *et al.*, the revolutionary FLINE model of planetary origins and evolution was brought full circle with the solution to the Fourth Law of Planetary Motion explaining the spacing of planets around our Sun. The original 12 orbits of our Solar System, along with Kepler's First Three Laws, verify both the dynamic origin of solar systems and the correct interpretation of Olbers in 1802 that Asteroids are fragments of disintegrated planets. Further, in view of the acknowledgment by astronomers that each giant gaseous planet in extrasolar systems is too close to its central star to have formed there via accretion of dust, gases and planetesimals, adds powerful evidence to the FLINE model that argues strongly against the accretion concept of planetary origins. Thus, this new model seems destined to displace the modified Kant-Laplace Accretion concept of planetary origins early in the 21st century. This change in direction of scientific thought will be expedited with the eventual confirmation of the true nature and sources of comets (ref. example, Comet SL9) and asteroids.

The inseparable connection between internal nucleosynthesis and evolution will lead to understanding the five stages of evolution of planets and moons, and why all active spheres are self-sustaining entities that create their own compositional matter. A prime example of the manner of creation of all planetary matter by means of natural laws is the origin and evolution of hydrocarbon fuels (erroneously called fossil fuels). With its foundation solidified by the Four Laws of Planetary Motion, the FLINE paradigm definitively connects past, present and future discoveries about all planetary systems. The recent discoveries about Mars, entering its fifth (inactive) stage, and other planets and moons continually fulfill predictions of the FLINE model, thereby enhancing its position as a valid scientific theory that warrants closer examination.

At this writing, the new Fourth Law explaining the spacing of planets has been in peer review by The Royal Astronomical Society in London for two years and four months.

References

1. A. A. Scarborough, The Spacing of Planets: The Solution to a 400-Year Mystery. Ander Publications, LaGrange, Georgia, USA, 1996.
2. W.C. Mitchell, The Cult of the Big Bang: Was There a Bang? Cosmic Sense Books, Carson City, Nevada, 1995.

Note: Such original works usually do not require or use many references other than commonly-known theories.

MORE ON THE SPACING OF PLANETS
History's Firm Foundation

In 1781 Sir William Herschel, while observing the constellation Gemini, noted that its brightest star was moving. It had to be a comet. Herschel could not believe that he had discovered an eighth planet; everybody knew that only seven planets existed in the Solar System (SS), seven being a magic number. Newton's laws enabled mathematicians to determine that the orbit was of a circular pattern, thus identifying the star as the eighth planet, Uranus. By the mid-nineteenth century, most astronomers had accepted this name, as suggested by the German, Johann E. Bode.

It was appropriate that Bode should have named the new planet, because he had predicted its existence in the correct orbit before its discovery. Bode had rediscovered a mathematical relationship first observed by Titius of Wittenberg in 1776. Director of the Berlin Observatory and author of a vast catalogue of star positions issued in 1801, Bode is best known simply for popularizing this mathematical relationship:

If to each number in the series 0, 3, 6, 12, 24, 48, 96, 192, one adds 4, then the series becomes 4, 7, 10, 16, 28, 52, 100, 196. When these numbers are divided by 10, the resulting figures are the approximate distances of the respective planets from the Sun: Mercury = 0.4 AU, Venus = 0.7 AU, Earth = 1.0 AU, Mars = 1.6 AU, Asteroids = 2.8 AU, Jupiter = 5.2 AU, Saturn = 10 AU. The general formula is $D = 0.4 + 0.1X$, where X is 0, 3, 6, 12, 24, 48, 96, 192, (the original series above).

This relationship is still known as Bode's Law (BL). At the time it was popularized, no planet was known to exist in the orbits at 2.8 AU and 19.6 AU. Herschel's discovery of Uranus in the orbit at 19.6 AU, as predicted by this law, spurred the search and discovery of Ceres in the orbit at 2.8 AU. However, when Leverrier discovered Neptune in 1846, it was found in a position much closer to the Sun than that predicted by BL. Since then, the significance of this relationship, or law, of the spacing of planets has remained a mystery wrapped in an enigma, conveniently discarded by many scientists as a quaint coincidence in numbers.

However, in true science, such data cannot be tossed aside only because its significance is not understood at the time. Although BL was not known in his time, Johannes Kepler (1574-1630), perhaps history's greatest astronomer, realized that the spacing of the six known planets was of a geometric nature. While successful in establishing the First Three Laws of Planetary Motion by solving their relevant mathematics, Kepler failed in his attempts to solve the geometric spacing of the planets – a potential Fourth Law of Planetary Motion.

The New Fourth Law of Planetary Motion

Had Kepler recognized the true significance of the precious jewel that actually was in use before 500 B.C. and that now seems destined to become the universal constant Phi, he might have utilized it as a basic tool for understanding the orbital velocity/distance/spacing relationship of each of the six known planets. When this was finally done in 1980 by the author, a smooth curve of the current (BL) Phi relationship of these properties of all planets was drawn (SS-6). Here, Phi would prove to be the key to understanding BL, and eventually to unraveling the mystery of the geometric origin of our SS. However, its full significance was not to be recognized until 1994-1995, some fifteen years later.

In 1994 the dawn began to break: The current spacing of planets could not possibly be their original orbits of some five billion or more years ago; everything in Nature forever changes; nothing remains forever as it once began; gravity and time exact their toll on all of Nature's systems. BL had to be simply a flawed version of the original perfect spacing of the planets (in accord with Nature's universal Phi). Anyone's original assumption that orbits remain forever stable was dead wrong – a stumbling block that had tripped too many theoreticians. The planetary orbits simply had been displaced from their original Phi orbits to their current positions, and will continue to be displaced by gravity forces throughout future eons. Thus, both past and future positions of the planetary orbits of our SS are calculable – a conclusion based on Phi mathematics and BL, and strongly supported by a plethora of scientific evidence.

In the concluding geometric solution (1995) to the spacing of the planets, the Phi figure (0.618034) represents the original distance of the closest planet (Mercury) from the Sun. The original distance of each planet thereafter was calculated by the formula, D = Phi x PO, in which the reciprocal of Phi is 1.618034 and PO is the distance (AU) of the previous orbit from the Sun. Moving in the reverse direction, Phi is 0.618034 and PO is the distance of the previous orbit (away from the Sun). Through utilization of Kepler's Third Law of Planetary Motion ($D^3 = T^2$), the original orbital velocity of each planet was calculated. These sets of Phi coordinates were plotted on the same GR scale used for the BL coordinates fifteen years earlier. The result was a curve almost identical to the 1980 BL-GR curve. Upon superimposing the two curves (BL and GR), certain conclusions became obvious:

1. BL is simply a flawed version of the original geometric GR (Phi) spacing of the planets.
2. The geometric spacing of planets is the crucial key to understanding the true scientific nature of our planetary origins.
3. The centuries-old enigmatic spacing of the planets finally has a provable solution based on mathematics, sound logic and strong supportive evidence.
4. *The gravity-induced displacement of any planetary orbit since the geometric origin of the SS is determined by the difference between its current (BL) position and its original Phi-spaced position. The original velocity of each planet is determined by Kepler's Third Law of Planetary Motion: ($D^3=T^2$).
   *(A plausible Fourth Law of Planetary Motion)
5. The future displacement, velocity changes and the fate of each planet of the SS is predictable mathematically.

Here was geometry at its beautiful best – beauty best defined by Richard Jefferies: "The hours when we are absorbed by beauty are the only hours when we really live…These are the only hours that absorb the soul and fill it with beauty. This is real life, and all else is illusion, or mere endurance."

With understanding, the concept grows in beauty and simplicity. During the past 28 years, much supportive evidence favoring this radically different perspective of planetary origins has accumulated in the scientific literature and in the author's files. Earlier in this book, three original sets of interlocking geometric diagrams, each set corroborating the other two and mingled in a pictorial history, explain how and why planets arrived at their current orbital positions in our SS. By combining this plausible Fourth Law of Planetary Motion with the author's previous writings on the evolution of planets via internal nucleosynthesis (the internal transition of energy into matter) in this revolutionary concept, reasonable solutions to a long list of anomalies of our SS become feasible.

The production of heavy elements comprising planetary crusts occurs in encapsulated energy masses as well as in supernovae and atomic bomb explosions. (Note: Heavy elements 99 and 100 were produced in the hydrogen bomb explosion). In this perspective, each planet is a self-sustaining entity: Via the natural processes of nucleosynthesis and polymerization, its mass of nuclear energy continually undergoes transformation into atomic elements that combine as molecules, comprising all matter of each planet as the sphere progresses through various stages of evolution from energy to gaseous to rocky to inactive sphere.

More on Geometric Origins

Binary star systems in which the two stars interact continuously are known to exist in large numbers. Imagine a potential binary star system in which the smaller, faster mass zips past the larger mass at a specific distance apart. If the two masses are of proper relative sizes and have proper relative velocities, the result will be a breakup of the smaller mass into fragments that split off at points determined by Nature's ubiquitous GR (Phi) geometry. With this special scenario in mind, one begins to realize <u>why and how</u> the planets of our SS were placed precisely in geometrically-spaced orbits. By combining this realization with the mysterious BL for the current spacing of planets, one can comprehend the plausible Fourth Law of Planetary Motion in which significant changes in the orbits and velocities of planets are revealed and future changes predictable.

The fiery origin of the SS is illustrated in the painting, *Birth of the SS* (Figure1). This artistic rendition of the manner in which the origin and geometric spacing of planets occurred allows an easier understanding of the geometry and the geometric forces illustrated in the SS diagrams. In it we see the breakout of Jupiter and its moons, far from the inner nebulous planetary masses orbiting the Sun in the lower left corner.

In every galaxy throughout the Universe untold numbers of binary star systems have been and are being created when two fiery masses, energized by Nature's more powerful explosions (Little Bangs), lock in orbit around each other. In <u>rare and special cases,</u> two such masses will meet the precise prerequisites for creating multiple planetary systems: proper relative sizes and velocities and proper distance apart to effect the fragmentary breakup of the smaller, faster mass as it swings by and beyond the larger mass <u>at a decelerating pace.</u>

In SS-6 the BL curve shows the current relationship between the orbital velocities and distances of planets when plotted in the Phi scale. What is the significance of BL, a formula for the geometric spacing of the first eight planets (including the Asteroids)? And why do the last two planets, Neptune and Pluto, fail to comply with this law? Is the spacing of planets really a crucial key to understanding and proving the origin of our SS? The answers to these questions reveal a beautiful continuity of evidence that offers golden opportunities to understand the origin and evolution of planets.

The nearly perfect AU-OV-BL curve SS-6, drawn in 1980, lulled the author into assuming this to be the original planetary orbits and velocities that had not changed since the original layout of the SS some five billion years ago. The assumption was wrong. Years passed before the author's realization that orbital positions and velocities do change, and that nothing can stay in orbit forever. The original positions obviously had to comply with Nature's ubiquitous Phi law of spacing, and the corresponding velocities had to comply with Kepler's Third Law of Planetary Motion! The BL curve for the current positions and velocities of planets is simply a flawed version of Nature's ubiquitous GR law of geometric spacing! A brief review of the solution to the Fourth Law, along with additional important information, appears warranted.

In July 1994, this original relative motion diagram, depicting the geometric origin of the SS, was modified to reflect a comparison of the BL data versus the GR data on the same geometric drawing utilizing the GR Triangle. Diagram SS-2 for the inner planets shows the relative motions of the Sun and the smaller mass during the origin of the SS. As the Sun moved at <u>a steady pace</u> to the right along the X-axis, the smaller energy mass (SEM) moved up the hypotenuse of the Architect Triangle (AT) at a <u>decelerating pace</u>, releasing a fiery fragment at each Phi point along the way. The original GR distance of 0.618 AU between the two properly proportioned masses was a crucial factor in the initial break-off from the SEM of a fiery fragment that was destined to evolve into Planet Mercury.

Diagram SS-1 is an expansion of the previous diagram to include the Phi release points and paths taken by all ten nebulous masses (including the Asteroids). For example, just as the Sun reached the right end of the X-axis, Pluto, the last of the ten masses, was pulled simultaneously into the final orbit at the top of the Y-axis. Had an observer ridden above the Sun during the time each nebulous planetary mass was pulled into orbit, the path of the smaller mass along the hypotenuse would have appeared to be the smoothly curved GR line extending from S (Sun) to P (Pluto), <u>an optical illusion caused by the relative motions</u> of the two masses.

The Architect Triangle is established via the standard geometric method of dividing a line into extreme and mean ratio. Contained within the Triangle are all the figures pertaining to the GR, the

Fibonacci Series, the Logarithmic Spiral, and perhaps to all the geometric forces of the Universe. The justification for its epithet was realized when its angles (26½° and 63½°) furnished a solid basis for construction of the seven SS Diagrams that intermeshed so precisely during the geometric solution to the origin of the spacing of the planets. In the author's opinion, a yet unknown connection exists between the AT and the super string theory of universal origins.

The warped curve identified as BL in SS-1 is a plot of the current positions of the planets. Thus, one can conclude that BL of spacing is simply a flawed version of the original GR law of geometric spacing of the planets—spacing warped by the forces of gravity over a very long time frame of some five billion years.

As the smaller mass swung at a 14^0 angle around the Sun, several other things happened. The passing SEM pushed the Sun's equatorial gases to a faster pace than its remaining mass and at an angle of 7° to the plane of the ecliptic. Thus began the erratic distribution of different degrees of inclination of the planets to the ecliptic, each effected by the combined forces of gravity of the Sun and of the preceding planetary mass(es). Mercury's angle of 7°, the same as that of the Sun's equatorial gases, is an important clue in verification of this concept.

At its Phi point, the Venus fragment broke from the upper part of the trailing edge of the rotating (counterclockwise) SEM, thus accounting, via a reverse gear-type action, for its slow, unique spin in the clockwise direction. Thereafter, at each successive Phi point along the way, gravity forces of the Sun and the previous planet(s) combined to pull the next fragmentary mass into orbit, each mass rotating in the counterclockwise direction. The layout procedure is best illustrated in the diagrams shown in SS-3 and SS-4. This second set of diagrams (modified versions of the original 1980 diagrams) shows <u>how</u> each planetary mass was pulled into orbit by the combined forces of gravity of the Sun and the preceding planet in accordance with Nature's ubiquitous GR. Note that no attempts to include the chaotic effects of the gravities of secondary planets in each case have been made.

In both diagrams, the Sun moves along the X-axis as the SEM moves up the Y-axis. The geometry of the forces of gravity pulling the <u>inner</u> planetary masses (SS-3) into their respective orbits furnishes corroborating evidence of the manner in which the clockwork precision described in Kepler's First Three Laws of Planetary Motion <u>came into being</u>. As each newly orbited fragment reached the tangent line (T or T`), its gravity acted in conjunction with the strong gravity of the Sun to pull the next mass into orbit, <u>thereby simultaneously varying the angle of inclination to the ecliptic for each planet-to-be.</u>

The second diagram (SS-4) of this set is expanded to show the geometry of the gravity forces acting on the outer planetary masses. As each mass reached the tangent line (T or T`) during the fiery layout of the SS, the next nebulous mass was pulled into orbit at a different angle of inclination to the ecliptic. If the original angle of 7° had been 0°, all planetary orbits would have been in the same plane.

In both SS-3 and SS-4, Y` can be made to coincide with Y at 90° by subtracting 14° from the baseline of 26½° to form a new baseline of 12½°. This shifts the exit points of all planets from Y` to Y and moves the T line down to T` to form a 39° angle with X. The significance of these moves that create a better-balanced Diagram is that the angle of diversion of the SEM around the Sun actually was only 14° instead of 26½°. While this figure is somewhat tenuous, it affects neither the conclusions drawn from the other Diagrams, nor the relationships among the sets of Diagrams.

The elliptical shapes of planetary orbits can result only from the rapid velocity of the smaller mass from which the nebulous planets were placed in their original Phi orbits. The multitude of known binary star systems and the scarcity of known planetary systems provide strong evidence that twin-star systems <u>very rarely meet the prerequisites of Nature's ubiquitous Phi geometry to become a SS like ours</u>. But whenever one does, energy fragments of a smaller, faster mass will break off at Nature's GR points along the way, all in compliance with the Phi geometry of the gravity forces acting at those points

The decision to draw the GR curve in the GR scale was made in early 1995 (see SS-5). Through use of the GR spacing of the planets and Kepler's Third Law, the original orbital velocities were calculated for every planet (including the Asteroids). Finally, the full dawn broke: the GR curve, when plotted in the

GR scale, reveals the original positions and velocities of the nebulous planets during the fiery birth of the SS. Since the 1995 GR (Phi) curve looked almost identical to the BL curve of 1980, why not superimpose the two curves to determine and evaluate the differences?

When the two curves are superimposed, they almost become one (SS-7). However, some significant differences are noted for each and every planet. Of the ten original orbits, eight have moved closer to the Sun, while Uranus and Neptune have moved slightly away. Changes in the velocities correlate with changes in orbital positions in compliance with the Third Law. Uranus and Neptune were the only two planets showing negative results: slight losses in velocity correlating with slight increases in distances from the Sun. NOTE: If a planet (or asteroidal matter) does exist beyond Pluto, its original distance and velocity calculate to 76.01AU and 1.006 miles per second, respectively.

Mercury, the innermost planet, has gained the most velocity (6.18 mi/sec) while being among the planets losing the highest percentage of its distance from the Sun. These increases in velocity and decreases in distance appear to be proportional to the mass/distance relationship of each planet and to the gravity effects of the preceding mass(es). Uranus and Neptune are examples of very distant planets, each with sufficient mass and distance to resist the inward pull of the combined gravities of the Sun and the inward planets; they are gradually drifting away while slowly decreasing in orbital velocity.

Contrary-wise, Pluto is so small that is has been, and still is, greatly affected by these gravities, in spite of its greater distance. This planet's tiny size permits it to be tossed around (relatively speaking) by any and all gravity forces within range, thus accounting for its highly irregular orbit and its decrease of 4.68 AU from its original Phi position at 46.98 AU. At this rate, Pluto's potential for collision with another planet in the far distant future seems real.

Besides Pluto, small significant differences between the superimposed curves occur at Uranus, Mars and the Asteroids. The large mass and distance of Uranus and the small mass of Mars account for their behavior, while the discrepancies of the Asteroids can be attributed to their vulnerable position, proportionately balanced between the two huge masses of Jupiter (the geometric mean of the SS) and the Sun. Apparently, the original Asteroids mass, broken into three main bodies, was pulled apart by the opposing gravity forces of the two larger masses. And exactly as Olbers concluded, the Asteroids disintegrated, exploded. Some 30,000 planetesimal remnants remain today in three distinct orbital paths, each beautifully spaced in precise geometric configuration in the SS.

From the superimposed curves, one can conclude once again that the BL spacing of planets is simply a flawed version of Nature's original GR spacing. This third set of diagrams correlates precisely with the results of the first pair of diagrams, showing the relative motions of the two masses of the potential binary star system during the original geometric layout of the SS some five billion years ago and the changes in planetary orbits that have occurred since then.

Developed during a time frame of 15 years (1980-1995), the seven geometric diagrams, each interlocking in corroborative support of the others, have come full circle to reveal the true beauty of the geometric origin of our SS. The enigmatic BL of spacing of the planets finally has been exposed as a flawed version of the original Phi spacing of the planets. In full compliance with the laws of thermodynamics, the fiery energy masses slowly evolve via the processes of nucleosynthesis and polymerization through various stages of evolution to end up eventually as cold, inactive spheres (see Chapter II).

From this pictorial history with its seven intertwined geometric diagrams, a plausible Fourth Law of Planetary Motion can be restated with a greater degree of confidence:

"The gravity-induced displacement of any planetary orbit since the geometric origin of the SS is determined by the difference between its current (BL) position and its original Phi position. The original velocity of each planet is determined by Kepler's Third Law of Planetary Motion ($D^3=T^2$)."

Any relevant anomaly can be explained better in this new FLINE concept than in any known theory of planetary origins. All are interwoven with a common thread throughout this revolutionary concept. While this book touches upon dozens of anomalies, many remain beyond the realm of its present objective. All of them offer numerous opportunities to pursue a true understanding of the origins and evolution of planetary and universal systems.

How Stable is the Solar System?

Questions have been raised about the stability of the SS. Is the particular distribution of the planets just one of many possible stable arrangements? Or is it the one arrangement that survives because it happens to result in a stable SS?

The Phi geometry of the original positions of the planets gives the impression of solid stability. However, when this Phi curve of the original orbits is superimposed on the curve of the current spacing of the planets, the answers to questions of SS stability become obvious. The chaotic displacement of the planets, while relatively small for all except Pluto, give a clear picture of the changes that have occurred since the origin of the SS. Each and every planet (including the Asteroids) has undergone changes in both its orbit and velocity. For example, the superimposed AU-OV curves (SS-7), reveal that Planet Earth has moved 38.2% closer to the Sun, while gaining 26.9% velocity. Similar changes for each and every planetary mass are recorded in the graph's tabulated data. From these changes, one can interpret the past and predict the future changes in each sphere's orbit.

The geometry clearly shows that there is no room to put another planet or two between the Sun and Pluto, the outermost planet. However, if one does not exist beyond Pluto, plenty of room is there for another planet — but perhaps insufficient gravity precludes that possibility.

Do Other Solar Systems Exist?

On June 11, 1994, the Hubble Space Telescope imaged the tiniest star ever recorded. Identified as Gliese 623b, the star is one-tenth as massive as the Sun and only one sixty-thousandth as bright. This diminutive mass lies 25 light-years from Earth, but shines too faintly and lies too close to a bigger, brighter companion (Gliese 623a) to be detected by ground-based telescopes. If placed as close to Earth as our Sun, the tiny star would have only eight times the brightness of a full moon. Astronomers estimate that the dim star resides some 200 million miles from its large companion and takes an estimated four years to orbit it.

The relationship between Gliese 623a and its tiny orbiting mass is governed precisely by the Four Laws of Planetary Motion. Astronomers know that the orbit has an elliptical shape and will sweep out equal areas in equal time while moving at accelerating and decelerating velocities in its elliptical orbit. In compliance with the Third Law, the cube of its distance from the larger mass will equal the square of its time of revolution. At this point in time, more information will be needed before the Fourth Law can reveal any displacement figures and the fate of the tiny star.

The establishment of the exact time of revolution enables scientists to calculate accurately the distance between the two masses and their distance from Earth. If the system is shown to be in compliance with the Four Laws of Planetary Motion, it might be a SS made in the same manner as ours. But unless smaller masses are found in other orbits, astronomers must conclude that the smaller mass failed to meet the precise specifications of the Phi geometry of the Fourth Law, and thus did not break up into smaller masses in the same manner as our planets did. Rather, the two masses took the usual route of remaining a binary star system, albeit an odd-couple type of system.

The dimness of the small star leads one to suspect it of being in the last stages as a star, and now entering the initial phase of the first stage of cooling and evolving into a gaseous planet in full accord with energy-to-matter laws; i.e., natural laws of physics, thermodynamics, chemistry, etc.

The big question then becomes: Is this the first recorded picture of a SS in its nebular stage of evolution from energy to planet? If so, it will rank among the most significant and exciting discoveries in the history of astronomy — one that can lead to a better understanding of our Universe.

The Ages of Stars and Galaxies

The conflict between the age of the oldest star versus the age of a younger Universe may force scientists to rethink the prevailing theory of the Universe. Observations by the Hubble Space Telescope imply that, under the standard theory, the Universe would be about 9.5 billion years old. But scientists are confident that the oldest stars are at least 12 billion years old.

The ages of stars have been extensively studied; they cannot be lowered without resorting to fundamental changes in scientific understanding of particle physics, according to astronomer, Nial Tanvir, of Cambridge University in England. Tanvir recently concluded that the standard assumptions about the cosmos might be wrong.

While such differences in ages cannot be understood within the realm of current beliefs about universal origins (the Big Bang), the problem simply does not exist in the perspective of the Little Bangs theory (1980) in which energy is continuously created at the spherical perimeter of the Universe during its speed-of-light expansion. Subsequently, this energy serves as the source for the evolution of new systems: quasars, galaxies, stars, planets, moons, comets, asteroids, etc., all evolving in full accord with natural laws, the Phi geometry and the mathematics of the Universe.

"When we try to pick out anything by itself, we find it is hitched to everything else in the Universe," stated John Muir, an astute observer indeed.

In December 1995, the Hubble Space Telescope focused for ten consecutive days on a tiny patch of sky near the handle of the Big Dipper, recording a brilliant picture that reaches far deeper into space than ever before. A bewildering array of galaxies was recorded from a composite of exposures from ultra-violet, blue, red and infrared emissions. The combined color image appears to show galaxies of all ages. In the current Big Bang perspective, the most distant ones were photographed as they looked when the Universe was only about one billion years old.

Two More Extrasolar Planets? (1996)

Since the writing of the first four chapters of the 1996 book, the discovery of two more extrasolar planets has appeared on the scene. At a meeting of the American Astronomical Society in San Antonio in January 1996, Geoffrey W. Marcy and R. Paul Butler presented a paper describing the discovery of two unseen planets, each orbiting a nearby star. One of the planets lies apparently at the right location from its parent star for liquid water to exist on its surface. The second planet might contain liquid water, but only in its atmosphere. (See figure 5)

The astronomers discovered the two planetary masses around sun-like stars – 70 Virginis in the constellation Virgo and 47 Ursae Majoris in the Ursa Major, also known as the Big Dipper. Although both stars are visible to the naked eye, the planetary masses are too small, and thus too faint, to be seen against the glare of their parent suns. So the researchers had to use the indirect technique of measuring small shifts in wavelengths of light emitted by the parent stars to find evidence of the existence of the orbiting masses.

Marcy and Butler monitored the motion of the stars with a spectrograph mounted on a 120-inch telescope. A computer analysis revealed that a light emitted by the two stars appears alternately redder and bluer, indicating that they move back and forth along the line of sight to Earth. In each case, the tell-tale wobbles describe a nearly perfect sine curve - a motion so periodic that only an unseen object pulling the star toward and away from Earth can account for it.

These discoveries, coming on the heels of the 51 Pegasi finding, are convincing evidence to scientists that planets are not rare at all. The planetary mass orbiting Ursae Majoris is about three and one-half times that of Jupiter. Circling its sun at about twice the distance of Earth from the Sun, the planet takes roughly three years to complete one revolution. Its surface temperature is estimated at -90°C.

In contrast, the sphere circling 70 Virginis has a mass about eight times that of Jupiter. Its orbit lies, on average, less than half Earth's distance from the Sun. Its surface temperature is about 83°C. Conceivably, the planetary sphere could have rain or bodies of water. However, assuming that it has a solid surface, its enormous gravity and high pressure would be crushing.

Further, the orbit of this planet is highly elliptical. Because of its strong gravity, a massive planet on an elliptical path tends to destabilize the orbits of nearby planets. Thus, the 70 Virginis is unlikely to possess an array of orbiting spheres similar to our SS.

There are other clues pointing to the origin and true nature of the two planetary spheres. First, the process of planet formation in which material accumulates from planetesimals and/or from a dusty disk rotating around a star does not permit a planet to have an elliptical orbit. Such elliptical orbits are permitted only in binary systems in which explosive momentum is created, as happens in the LB/FLINE concept. (See Chapter V.)

Second, the 51 Pegasi planetary sphere is much too close to its Sun to have formed from a gaseous dust-cloud that either would have burned away or fallen into its parent sun. Further, it is difficult to imagine such a fast-moving target being bombarded with in-falling planetesimals that add substance and heat to its mass, as is hypothesized for planets of our SS, a vital process for building planets in the perspective of the BB theory.

To understand the true origin and nature of the three new solar systems, one must consider the relationship discussed previously: the relative sizes of each pair, their relative distance apart and their relative velocities. In this perspective, the logical conclusion is that all three of the new systems are representative of peculiar Mutt-and-Jeff pairings of binary star systems.

The large sizes of the two new planetary masses, each circling its parent star, indicate that they are either in the first stage (energy masses) or the second stage (gaseous) of evolution. Because of their extra large sizes, any betting odds would have to favor the first stage of evolution. The smaller 51 Pegasi mass, being only 0.6 the mass of Jupiter, could well be in the rocky stage of evolution. This statement is based on the assumption that it is no younger than our SS planets and perhaps somewhat older.

All three are fine examples of solar systems that failed to meet the precise prerequisites of the Phi geometry that proved so essential in the geometric spacing of the planets of our SS. Thus, it appears that each of the three new discoveries is destined to remain a binary system rather than a geometrically-spaced multiple-planet SS like ours. (For relative locations of planets, see Figure 5.)

The full significance of the three discoveries may not reach fruition for many years. While presenting exciting evidence supportive of the new LB/FLINE concept, they argue strongly against the Accretion Disk hypothesis, and subsequently against its well-entrenched ally: the Big Bang theory.

Introduction to the Five Stages of Planetary Evolution (1)

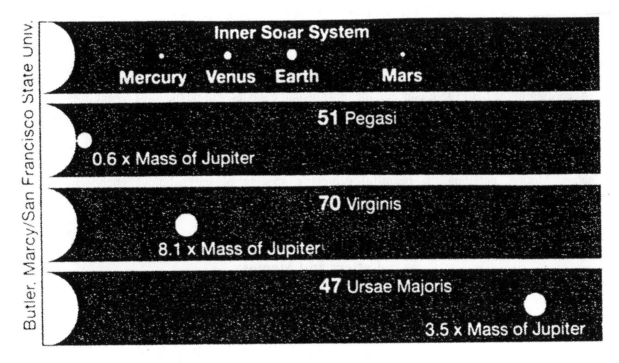

The relative locations of planets in our Solar System and the newly discovered planets orbiting 51 Pegasi, 70 Virginis, and 47 Urae Majoris.

Figure 5

Extrasolar Systems: How and Why They Differ From Our Solar System.

The Four Laws of Planetary Motion reveal the absolute necessity of a great momentum of a smaller, faster mass interacting with a larger mass as it sped past our Sun to create the elliptical planetary orbits with geometric clockwork precision. Further, the solution to the Fourth Law reveals that in order to create a multiple-planets solar system like ours, in which the smaller planetary masses break from the speeding mother mass at the GR points along the way, the two initial masses must meet precise prerequisites of relative velocities, relative masses, and a specific distance apart to effect a critical angle of diversion and subsequent breakup around a central star. Since it is a rarity for two masses to meet these exact prerequisites, the vast majority of solar systems will contain only one intact giant planetary mass circling its central star.

At this time (Dec., 2000) astronomers have discovered about 50 of these exoplanetary solar systems, all at very great distances from our SS, and each with a planetary mass that failed to meet the prerequisites specified in the solution to the Fourth Law of Planetary Motion. As astronomers have acknowledged, each of these giant gaseous planets is much too close to its central star to have formed there via the prevailing accretion concept of planetary origins!

All solar systems comply with Kepler's First Three Laws. Whether the solution to the Fourth Law applies solely to our Solar System remains to be seen. But as we shall see, the sizes and gaseous nature of all giant Jupiter-like planets, as well as the sizes and nature of all smaller planets and moons in our Solar

System, can be explained by a new concept of planetary evolution that is based on the Four Laws of Planetary Motion.

About 1,500 galaxies, many only one four-billionth as bright as the dimmest light the unaided human eye can see, can be seen in the spectacular Deep-Field picture. The galaxies are stacked up against one another, so the real challenge is to disentangle them. A high priority is to determine the distances from Earth for as many of the galaxies as possible. Since distant galaxies move away from each other faster than nearby galaxies, astronomers can determine within reason each one's distance from Earth.

According to the LB/FLINE concept, the most distant galaxies will be the most recent ones created at the expanding perimeter of the spherical Universe, and will be moving at nearly the speed of light (details in Chapter V).

Thus to fathom universal and planetary origins, mankind must learn their mathematical relationships and their continuity in the chain of evolution from energy to all forms of matter. The new Fourth Law of Planetary Motion, detailing the beautiful geometric spacing of the nebulous energy masses that subsequently and continuously evolve via nucleosynthesis and polymerization as self-sustaining planetary entities, brings this conclusion sharply into focus.

Letters and Memos

To: Letters Editor, Science Magazine, January 26, 1998

In writing about the problem of how the giant gas planets got so close to a central star (*Migrating Planets, Science*, Vol.279, p. 69) Murray *et al.* based their findings and conclusions on the assumption that there is no known formation method that would allow such large planets to initially form that close. They suggest that a giant planet forms in the outer parts of its system and then migrates toward the star, pushed via instabilities created by planetesimals that linger in the disk after most of the gas has been consumed in planet formation.

Two counterpoints are warranted here: First, the formation of any singular or multi-planet solar system must comply with the basic principles of the Four Laws of Planetary Motion. These principles did not exist in the article.

Second, the assumption that there is no known formation method that would allow such extrasolar planets to initially form so close to the central star is not an accurate statement. A formation method that also complies with the basic principles of the Four Laws of Planetary Motion does exist, although it is not yet well known.

Kepler's First Three Laws of Planetary Motion are familiar textbook material. Together, with the recent mathematical solution to the proposed Fourth Law of Planetary Motion (1995) governing the geometric spacing of multi-planets in solar systems, the Four Laws provide sufficient clues for solving the mystery of how gas giants got so close to the central star. This evidence provides a valid scientific alternative to the findings described in the *Migrating Planets* article. Meanwhile, the findings on Mars and of other space probes continue to corroborate the only other choice (the FL/IN/E paradigm).

To: Christine Gilbert, Letters Editor, SCIENCE, Washington, DC

Thank you for the copy of Dr. Norman Murray's response of February 26 to my letter of reply to his paper *Migrating Planets (Science* Vol.279 p.69). His frank response touches on some of the problems that scientists face en route to the truth about planetary origins and evolution.

Kepler's First Law of Planetary Motion deals with ellipses rather than "the radius vector of each planet with respect to the sun sweeps over equal areas in equal time," (the Second Law). But Murray is correct in stating that "it is unlikely that Jupiter formed at 0.05 AU...Jupiter's orbit has shrunk." Actually, the Four Laws of Planetary Motion clearly reveal that Jupiter formed initially at 6.854 AU and now resides in the 5.39 AU orbit (an inward movement of only 1.46 AU in some 5-6 billion years). Perhaps he meant the Jupiter-size mass in an exoplanetary system.

While these small mistakes can be overlooked, Murray's letter gave the alarming impression that he is unaware of the new Fourth Law of Planetary Motion detailing how the planets of our Solar System (SS) attained their current orbital positions around the Sun — the law that first eluded Kepler in 1595. Further, the Fourth Law explains (1) the unorthodox orbits of Neptune and Pluto, (2) the outward drift of Uranus and Neptune, and (3) the ecliptic angles, etc. Together, the Four Laws clearly reveal how our SS came into being, while clearly disproving the condensation/accretion hypothesis of planetary origins from a planetesimal disk (which accounts for the difficulties in getting it past peer reviewers for publication and for the impossibility experienced by Wetherill in attempts to correctly space the planets via computer trials based thereon).

Thus, the Four Laws make it unnecessary to ponder whether large planets can or cannot form so close to their parent stars via the prevailing hypothesis. But clearly their orbits (original and current) can be any distance from the parent star as long as they obey the Four Laws of Planetary Motion. The reason resides in the fact that the distance is a function of relative sizes, original velocities and initial distance apart to effect the end result of being in close, medium or distant orbits around the parent star. This information is the basis of my argument against Murray's statement "that there is no known formation method that would allow such large planets to initially form that close." Although little known and less publicized, this provable formation method does exist in the scientific literature - in my latest book, *THE SPACING OF PLANETS: The Solution to a 400-Year Mystery*. A paper on the subject is again in process of peer review for publication; however, its irrefutable arguments against prevailing beliefs about planetary origins might again prove too revolutionary and too unsettling for publication at this time.

I argue against only the basic foundation and conclusion of Murray's math - not his math. Evolution (E), Internal Nucleosynthesis (IN) and the Four Laws of Planetary Motion (FL) are inseparable: one cannot exist in any SS without the other two. If we are to understand all anomalies of solar system and planetary evolution, we must first understand the relationship among these three tangible facets of creation that are the basic foundation of the provable FL/IN/E paradigm of planetary origins and evolution.

Thank you again for giving us the opportunity to discuss our differences.

To: Glenn Straight, Science Editor, The World & I, Washington, DC 20002. 06-02-98

The enclosed article *Images emerge of suspected planet 450 light-years away* (Atlanta Journal-Constitution, May 29, 1998) contains some crucial information pertaining to planetary origins - information predicted and explainable via my proposed FLINE paradigm. To quote Toner's article, "If the newly discovered object, named TMR-1C is confirmed as a planet, it would force scientists to revise their notions that giant planets like Jupiter require millions of years to coalesce out of a solar, or stellar dust cloud." Such confirmation and revised notion would indeed add powerful corroborative evidence to my writings of the past 25 years on planetary origins and evolution. It is a giant step in the right direction: planets do not coalesce from gaseous dust clouds—its speed of more than 20,000 miles per hour and its trail of glowing gases are crucial factors that cannot be ignored as clues to a definitive understanding of its origin and nature.

Once past this roadblock, the next advance in planetary origins will come with the acknowledgement that planets do indeed evolve through five stages of evolution: from energy to gaseous to transitional to rocky to inactive, or dead spheres. Without the original energy core mass that drives each planet through these five stages before its depletion, evolution simply would not be possible. Kepler's Three Laws of

Planetary Motion and the proposed Fourth Law of Planetary Motion explain how the original energy masses were placed in geometrically-spaced orbits around the Sun before undergoing evolution via Internal Nucleosynthesis into the planets we observe today - all in accordance with size and in full compliance with all natural laws. Thus, in our Solar System, <u>the Four Laws of Planetary Motion (FL), Internal Nucleosynthesis (IN), and Evolution (E) are inseparable: one cannot exist without the other two.</u>

The article touched on an important key to understanding planetary origins: ubiquitous binary star systems. The geometric solution to the Fourth Law of Planetary Motion reveals that our Solar System is a very special and rare situation in which the two stars of a potential binary system met the specific prerequisites of relative size, relative velocities and distance apart to cause the smaller, faster mass to break up into fiery fragments that were pulled into orbit in full accord with nature's ubiquitous Phi (or Golden Ratio) geometry. The Seven Geometric Diagrams detail this event beyond the shadow of a doubt.

The newly discovered planet is another example of a potential binary star system that failed to materialize in the usual manner: a "planetary nomad, cast adrift in interstellar space by a newly forming pair of binary stars." It will eventually evolve through the five stages of evolution common to all planets. Binary star systems and planets originate from powerful explosions (perhaps of quasars or black holes) that scatter the fiery masses along their rapid lines of flight (a facet of the Little Bangs Theory that is beyond the scope of this letter, but is detailed in my book, *THE SPACING OF PLANETS: The Solution to a 400-year Mystery*).

This important discovery should convince most skeptics that the time indeed has come to revise our notions about the origin and evolution of self-sustaining planets.

To: Glenn Strait, Science Editor The World & I Washington, DC 20002,   4-20-99

Thanks for the news release about the discovery of the first multiple planet system around a nearby Sun-like star, which was already known to have one planetary companion. The news had come off the AP wire one day after my return from the National Philosophy Alliance (NPA) meeting in Santa Fe. Had it been known two days earlier, I would have used it as additional evidence corroborating the FLINE paradigm in my paper there. The planetary data fit precisely into the new concept — as will that of any system with any number of planets. The sizes, velocities and gaseous nature of the three huge Jupiter-like planets account for their positions in the system and their being in the second stage of the five common stages of planetary evolution. Simply put, the FLINE paradigm answers the big question of how such a solar system arose.

Marcy's statement that "no theory predicted that so many huge planets would form around a star" fortunately is not true. Apparently, he is unaware of the FLINE paradigm that both predicted and explains this anomalous situation. It is frustrating to read such erroneous statements while remaining unable to get this viable alternative into the scientific community through normal peer review channels. As long as the planetary accretion concept prevails, erroneous speculation about planetary origins will persist.

Immediately after my presentation to the group of some three dozen international NPA members and visitors, I asked Dr. Roland Hron if he had been able to hear it clearly. His response: "Yes, and it was spellbinding." To make the paradigm crystal clear, I had used 23 transparency illustrations to effect its full message. Apparently, its factual, non-speculative, common sense nature makes this concept easy to understand and accept whenever it is given the opportunity to be heard in its entirety. If you would like a copy of my paper, *The FLINE Paradigm: Definitive Insights Into the Origins, Orbital Spacing and Evolution of Planets* featuring the proposed Fourth Law of Planetary Motion, I would be happy to send a fully illustrated copy for editing and publication.

P.S. Dr. George Galeczki (President, Society for the Advancement of Physics) Cologne, Germany, informed me of similar work on planetary spacing being done by Beckmann. He promised to send a copy of the book as soon as it is translated from German into English. Galeczki has one of my books.

Recent Corroborative Evidence

The Extrasolar Planet With an Earthlike Orbit (Science News, Aug 14, 1999) brings up a good point. Of some 20 known planets outside our solar system, the latest find, announced July 29, 1999, stands out from the group. The planet's mass is a least 2.26 times that of Jupiter, and circles the sun-like star, iota Horologii, at a distance that's 29 million km (20% of the Earth-Sun distance) from their parent star. Its orbit is elliptical. To quote astronomer Geoffrey W. Marcy: "This new planet adds to the suspicion that our solar system with its neat, circular, coplanar orbits, may be the exception rather than the rule."

Definitive reasons for our solar system's being the exception rather than the rule are clearly stated in the FLINE model (1973-1997) of planetary origins and evolution. The geometric solution to the new Fourth Law of Planetary Motion (1980-1995) (in conjunction with Kepler's First Three Laws) explains how our planets attained their orbital spacing around the Sun. In the vast majority of instances in which two fiery masses interact, the results will be a binary star system, as described in *The Spacing of Planets: The Solution to a 400-Year Mystery* (p. 12). If the two masses are of proper relative sizes and have proper relative velocities, and the smaller, faster mass zips past the larger mass at a very specific distance from it, the result will be a multiple-planets solar system like ours—truly a very rare event!

When these very specific conditions are not met, as happens in the vast majority of cases, the result will be simply another of many solar systems consisting of one or two extra solar planets. In all cases, the planets in question eventually evolve through the five common stages of evolution. As a planet gravitates ever closer toward its central star, its orbit should become less and less elliptical.

Marcy's suspicion that our solar system may be the exception rather than the rule is in full agreement with the FLINE model's predictions and explanation of why the event is so rare. To go one step further, the solutions to all solar system and planetary anomalies are deep-rooted in the three basic principles of this paradigm. Only in the FLINE model can fully provable solutions to all anomalies of solar and planetary systems be realized.

Our SS's planetary orbits are somewhat chaotic, elliptical, and multi-planar, just as is revealed by a plethora of evidence; they definitely are not "neat, circular, co-planar orbits!"

Plentiful Planets (Science 17 December 1999, p2241)

The article discusses the extrasolar planets discovered during the past 4 years: "All planets detected so far are gas giants. Moreover, most of them move in very eccentric orbits, unlike the planets in our solar system, leaving theorists scrambling to explain why. Although the string of new data confirms that planets are common, at this stage it seems that planetary systems configured like our own are rare indeed... In 1999 astronomers caught an exoplanet transiting the face of its star, significantly dimming that star's brightness for a few hours every 3.5 days. It is 200 times more massive than Earth, has a diameter significantly larger than Jupiter's, and like Jupiter, is composed mainly of hydrogen and helium."

All the data compiled from these important discoveries were accurately predicted and are readily explained by the new definitive FLINE model. The eccentric orbits of the larger-than-Jupiter planets, their huge sizes and low densities are critical clues revealing the dynamic origins of the systems whose planets, because of their giant sizes, are still in the early phases of the second (gaseous) stage of their planetary evolution, perhaps a billion years or so behind Jupiter's current second stage of evolution.

Since binary star systems and the extrasolar systems are created in an ongoing and similar dynamic manner, astronomers can expect to find many more extrasolar planets and very many more binary star systems. The major differences between the two systems, as with our own planets, can be accounted for by the size differences of the masses in question.

The reason for the rarity of a solar system like ours, as predicted and explained by the FLINE model, resides in the fact that the dynamic layout of our system was accomplished under exacting specifications that are in full accord with the Golden Ratio. These specifications call for two energy masses of precise relative sizes and relative velocities, both masses interacting at a precise distance apart to effect a layout in full accord with the formula D=1.618034 x Dpo, where D is the distance in AU of any pair of orbits, and Dpo is the distance in AU between the two previous planetary orbits. These specs and formula are the keys to the solution to the new Fourth Law of Planetary Motion detailing how the planets attained their orbital positions around our Sun. All of the original 12 nebula planetary masses (including the original 3 asteroids spheres that later disintegrated, as correctly interpreted by Olbers in 1802) adhered strictly to Kepler's First Three Laws of Planetary Motion as well as to the new Fourth Law.

The short article, *Flat and Happy*, on the same page of *Science* describes how the theory of inflation "holds that a burst of expansion in the first instant of time stretched space almost perfectly flat, so that parallel light rays stay parallel forever. Measurements of distant exploding stars suggested that a mysterious energy in empty space is accelerating the expansion of the universe." Astronomers use the tiny "fluctuations in the microwave background as proof that energy can curve space just as matter can, and the finding suggested that energy might fill in for matter and flatten the universe." Surely, Occam's razor can be utilized here to render a more logical explanation.

Einstein was correct in two relevant statements: (i) space is curved, and (ii) his earlier statement that an unknown force (lambda) is acting against the inward forces of the universe was the biggest blunder of his life. When viewed in the perspective of the Little Bangs theory, space is indeed curved and is expanding at its perimeter at the speed of light; no lambda factor is needed to explain the naturally accelerating expansion of the spherical universe. In the vacuum of space everything is spherically shaped, including the universe itself. Why must everything become more and more complicated via the speculations of a Big Bang origin? When will the tangled web cease to be woven?

Shaking Up a Nursery of Giant Planets. (Science, 10 Dec 1999)

Quoting from the article: "What are Uranus and Neptune doing so far from the sun? The question has puzzled theorists for decades. Unlike the closest six planets, which orbit the sun inside of 10 astronomical units (AU)— that is, less than 10 times farther out than Earth—Uranus and Neptune orbit at 19 and 30 AU, respectively. Theorists don't believe they could have formed so far out; there, gas and dust were too sparse to coalesce into planets. Now a new computer model suggests that sibling rivalry might be to blame for their banishment. Runty Uranus and Neptune may have grown up in tight quarters much closer to the sun, only to have the big bruisers, Jupiter and Saturn, fling them into the outer reaches of the solar system."

The reason for this logic? "...out on the nebula's fringes, matter was spread too thinly for anything like planets to form. In the best simulations of the process, cores for Uranus and Neptune fail to form at their present positions in even 4.5 billion years, the lifetime of the solar system." To quote one astronomer: "Things just grow too slowly in the outermost solar system. We've tried to form Uranus and Neptune at their present locations and failed miserably." So the astronomers theorize that the two planets did form closer in where the nebula was far denser, and then were flung into the outer reaches of the solar system.

A number of years ago, other computer simulations failed miserably in efforts to create planets properly spaced in their current orbital positions. The source of the problems leading to these (and future) failings resides in the Accretion hypothesis itself; planets did not accrete from gas, dust, planetesimals, comets, etc. There is a much simpler and more viable alternative that poses none of these problems with our Solar System or with an extrasolar system - a revolutionary definitive concept structured with five basic and irrefutable principles of Nature.

The solution (1980-1995) to the enigmatic Fourth Law of Planetary Motion clearly reveals how the planets were placed in their orbital positions around the sun during the layout of our solar system some 5 billion years ago. The Four Laws of Planetary Motion (FL) clearly support a dynamic origin of our solar system in which the nebulous planetary energy masses (first identified by Descartes in 1644) have evolved via internal nucleosynthesis (IN) (in compliance with Einstein's formula $E=mc^2$) into their current stages of a five-stage evolution (E) common to all such spheres throughout the Universe. We have only to look to the sun and planets of our own system to observe these five stages of planetary evolution. The three principles embodied in the FLINE model are inseparable and ongoing until each sphere's energy core is depleted and the planet (or moon) has entered its inactive fifth stage of evolution.

The fact that Pierre Simon Laplace's Accretion hypothesis of 1796 (with some modification) remains the dominant concept of planetary origins in spite of its highly speculative structure and much contradictory evidence, and in the face of even stronger evidence favoring the FLINE model, speaks volumes about the status of our planetary sciences. With two notable series of computer failures of recent times, one must wonder how much longer this antiquated hypothesis can survive the onslaught of opposing facts, and how much longer it will serve as a costly roadblock to understanding the full significance of the brilliant past, present and future discoveries in outer space.

In the absence of the FLINE model, the web being woven around the Accretion hypothesis becomes more and more tangled. In the new perspective the solution to every anomaly of every planet and moon has its taproot deeply embedded in the FLINE model - a statement already proven true by its splendid record of accurate predictions.

A Field Guide to the New Planets

A Field Guide to the New Planets (Discover, March 2000) is an interesting article on the 29 new planets discovered orbiting stars like our own with pictorial data concerning the 29 gaseous giants. The article states: "the orbits in which [our planets] were born some 4.6 billion years ago have remained the same ever since. Until recently that was the accepted scenario. But now the detection of extrasolar planets has forced astronomers to re-examine such notions, because they present us with a paradox. Many are so monstrous in size, and hug their stars so closely, that they could not have formed in their present position. The searingly hot stars around which they circle would have melted their rocky cores before they got started. Instead, it's assumed that they coalesced some distance away, then barreled inward over millions of years. And if such chaos characterizes the birth of extrasolar planets, could not similar disorder have reigned closer to home?" — A proposed scenario that presents more problems than it solves.

However, there is an alternative that warrants consideration; one in which the data does fit precisely and without need of the assumptions of the proposed scenario. Such extrasolar planets were predicted and explained by this revolutionary concept, even before these amazing discoveries were made. In it, there is no need for such disorder in solar systems; they are precise systems that adhere to the Four Laws of Planetary Motion (FL) and the processes of internal nucleosynthesis (IN) that drive the evolution (E) of all active celestial spheres. These gaseous giants are simply in the early part of the second stage of the five-stage evolution common to all planets - all in full accord with size and the laws of thermodynamics. The revolutionary FLINE concept eliminates both the need for gaseous giants to form at much greater distances from their central stars and the need to sling planets out of their orbits in a chaotic manner.

The new Fourth Law of Planetary Motion that brought the FLINE model full circle explains how the planets attained their orbital positions around our sun - an enigma that first eluded Kepler in 1595. And we have only to look to the skies to observe the five common stages of planetary evolution.

To: Sue Bowler, Editor, The Journal of the Royal Astronomical Society, UK. Aug '99

On the Spacing of Planets: The Proposed Fourth Law of Planetary Motion. Ref qyO16; Submitted January, 1998.

Thank you for your letter of 7 July concerning the above paper and the apology for the long delay in the hands of the referees. At that time you anticipated knowing the outcome of the refereeing process within the next two weeks.

If our current correspondence hasn't just passed in the mail, please advise the status of the paper and whether there are questions about any unclear points in the article and its illustrations.

Conservatively speaking, the significance of the Fourth Law appears equal to that of Kepler's First Three Laws of Planetary Motion, while its revolutionary impact on the sciences of solar systems should equal that of the Copernican idea of a Sun-centered Solar System. Certainly, it does hold much promise as the crucial key to understanding and proving the origins of solar systems. As such, and in the interest of science, I'm sure you agree that its evaluation and publication warrant a high priority.

Thank you for your valuable time and kind consideration.

Chapter II

HOW PLANETS EVOLVE

The FLINE Paradigm: Definitive Insights Into the Origins of Solar Systems and the Orbital Spacing and Evolution of Planets.

Abstract

Three fundamental and inseparable realities of nature comprise a revolutionary paradigm of planetary origins and evolution The first two realities are of a geological nature: planetary evolution (E) and internal nucleosynthesis (IN). In any sphere, neither E nor IN can exist without the other. These realities provide the connecting links among all the discoveries and mysteries of Earth's surface features and its internal characteristics. The same IN and E principles are applied to the evolutionary cycles and characteristics of all planets and moons. The critical connection is made between planetary evolution and the essential source of energy that drives planets through five stages of evolution common to all such spheres until each planetary core is depleted of energy, leaving only an inactive (dead) sphere (e.g., Mercury) in orbit. The supporting evidence embraces thirteen clues to the true nature of planetary cores: Sun-like energy, the taproot of the mechanisms of magma replenishment into caldera magmatic systems and other fiery outlets. This paradigm provides an understanding of the role of the inner core in Earth processes, including the generation of the geomagnetic field and the thermal evolution of Earth. The internal nucleosynthesis-evolution (IN/E) concept leads inevitably to the mathematical solution to the proposed Fourth Law of Planetary Motion (1980-1995) that first eluded Kepler in 1595. Together with Kepler's First Three Laws of Planetary Motion, the Four Laws (FL) (the third fundamental reality) bring the FL/IN/E paradigm full circle by definitively revealing how nebulous planetary masses, including exoplanets, attain their orbital spacing around their Sun. Prevailing theories fail to make these connections between SS origins, orbital spacings and the evolution of planets, moons, etc. Discoveries of the space probes to Mars, Venus, Europa, Jupiter, etc. continually add corroborating evidence.

Introduction to Planetary Evolution

All planets began as relatively small masses of fiery energy: small stars, each placed in orbit in full accord with the Laws of Planetary Motion. Each is destined to obey these Laws as long as it remains in orbit around its sun.

The transition of our planets from nuclear energy to matter has required some five billion years of evolution to reach the current stages observed in the nine planets and the asteroids of our SS.

The Phi geometry of the SS in Chapter I revealed how and where each planet was placed in orbit, and how much each has been displaced from its original orbit. Knowing this, we can uncover many clues that divulge the secrets of how each planet evolved by means of natural laws to its present stage of evolution. This is possible within the realm of the natural laws of physics and chemistry, and can be accomplished without speculation or assumptions that too often mislead.

How is it possible for planets and moons to evolve from nuclear fireballs into the spheres of the SS as observed today in their geometrically spaced orbits? Our Sun furnishes significant clues to the evolution of planets by natural laws. An atomic forge, it is readily recognized as one in which fission-fusion reactions release energies in the form of heat and light from atomic nuclei undergoing constant changes. For example, four hydrogen nuclei fuse together to create one nuclei of helium, and almost 1% of the original weight or mass of the hydrogen material is changed into heat and light.

The glowing hot gases of our Sun are made of the lighter elements that comprise the huge gaseous planets and are found in the crust, oceans, and atmosphere of Earth. The most common elements in the Sun are hydrogen, helium, calcium, sodium, magnesium, and iron; the number of other elements includes oxygen, nitrogen and carbon. The main point here is that atoms are created in situ from the energy of nuclear masses under extreme conditions of temperature and pressure. The big question becomes: How does a nuclear mass (a small sun) metamorphose into a planet or moon?

The first phase of transformation (see Figure 3, Chapter I) begins with the accumulation of countless numbers of atoms, ions and simple compounds that eventually form an enveloping spherical blanket high above their source (the energy mass, now identifiable as the core). Simultaneously, similar matter, as with the Sun, accumulates on the extremely hot "surface" of the core. During billions of years, virgin matters pours from the manufacturing plant below, slowly filling the huge space between the energy mass and its spherical blanket to form clouds of chemical vapors throughout the towering atmosphere. Scientists see the results in Jupiter and Saturn, both in the second stage of planetary evolution, and classify them as gaseous planets.

In the second phase of transition, the voluminous clouds of towering atmospheric matter gradually close in on the hot interface, and precipitation of its chemical vapors is initiated. The precipitates are instantly repelled by heat; the gigantic battles between the evaporation and the condensation forces rage on for eons. The atmospheric matter is trapped between the unbearable heat of the energy mass and the super coldness of outer space. In time, matter gains a toehold in the form of the first tenuous liquid layers, eventually followed by thin, tenuous crustal formations. Uranus and Neptune are into this transitional third stage of evolution, although their outward appearances may or may not justify their classification as gaseous planets.

During the third phase in the evolution of a planetary sphere, shallow seas cover the entire surface, cooling it sufficiently to permit formation of a more permanent crust. The crust thickens as a function of time, a continual process for billions of years. Newly created matter from volcanic outpourings of lava, elements, water, gases, and compounds of great varieties continually build and alter atmospheric and land systems. Planet Earth is an excellent example of this fourth (rocky) stage of planetary evolution.

With these dramatic, often violent, and persistent environmental changes, species, where feasible, have their beginnings; they come, they flourish, and they vanish into extinction as the ever-changing environment dictates. They adapt or they perish. The era of the dinosaurs is a prime example – perhaps the most popular of all times. Their history offers powerful evidence that Earth is a self-sustaining entity creating its own systems from within, always maintaining control of the creation and extinction of its species. Other prime examples include the life that exists around the "black smoker" vents discovered on deep ocean floors, along with outpourings of virgin materials that never cease. Even more voluminous outpourings of virgin materials occur from the baseball-like seam circling the globe, as often viewed on TV screens.

There is no need to look to outer space for answers to the origin and extinction of species. As it continuously thickens over time, Earth's multi-layered crust offers clues to events effected by virgin ejecta, including the iridium layer at the K-T boundary. Although other layers of iridium have been found in other eras, this particular iridium layer has been interpreted erroneously as coming from outer space and wiping out the dinosaurs. In reality, it is a layer of ejecta matter created within Earth via nucleosynthesis under extreme, specific conditions of temperature and pressure – as were the other iridium layers.

A similar example of a puzzling heavy metal layer was discovered recently on Venus (*Science,* 5 Jan 1996). The loftiest parts of the planet, like the highest peaks on Earth, are covered by a perpetual frost – a coating of the lustrous, silvery white element tellurium. This explains why the highlands of Venus appear so bright in radar images of the surface. Tellurium has just the right electromagnetic properties to explain the radar brightness of Venus, and just the right melting point to coat its highlands, but not its plains.

Planetary scientists now argue that the metal, nearly as rare as gold, could be spewing from volcanoes as metallic vapors that would freeze out onto the cooler highlands. This is in full agreement with the 1973 idea that has developed into the FLINE concept during the past 28 years, and is powerful confirmatory evidence that the iridium layers on Earth were made is the same manner via the nucleosynthesis that exists in both (and all) planets. Also, it is powerful evidence of the ever-changing planetary environments that spawn and eventually kill the species that cannot adapt to these changes: the true cause of the demise of the dinosaurs.

As planets evolve from tiny, fiery stars into crusty old spheres with ever-diminishing nuclear cores, startling things happen. In Earth's case, the hot, humid environment spawned the first small wonders of life. Bacterial life spawned in temperatures of a few hundred degrees before the surface cooled sufficiently to allow the formation of eubaryotic cells. It was only a matter of time and additional cooling before various forms of life could evolve and be sustained.

Dinosaurs: The Reason for Their Extinction

Eventually, Earth's average temperature cooled to the mid-100s, with high humidities, heavy rains and vicious thunderstorms prevailing: ideal conditions for plush tropical forests worldwide. The steamy forests were fully capable of sustaining huge life forms – a real paradise for giant animal life. Dinosaurs thrived on the lush plant life and smaller animals during the 140 million jungle years of the Mesozoic Era, often referred to as the Age of the Dinosaurs.

During this rich time span, environmental conditions were changing ever so imperceptibly. Temperatures gradually declined as Earth's crust thickened, its insulation propensities ever-increasing to block out the heat emanating from within. The food supplies of Nature's lush forest began to diminish gradually, while the high oxygen content of the atmosphere declined proportionally over the ages.

Vicious, fiery eruptions from Earth's hot interior were commonplace – the order of the day – as evidenced by the multiple layers of ejected materials that together with sedimentary layers comprise the crust. With each passing millennium, the composition of the erupting materials changed in accordance with the prevailing internal conditions of both temperature and pressure as functions of time that determined the type of end products added as new crust. Each new layer of material contributed its effects, good or bad, to the changing surface and atmosphere. These crustal layers now serve as clocks and books from which mankind extracts their recorded history.

With the passing of time, the dinosaurs adapted as best as they could, while slowly declining in numbers as the environment became ever more hostile. Gradually, Nature began delivering its coup de grâce from the bowels of the mother Earth that had spawned and nurtured the giants through almost 140 million years, and now was dictating their rapid demise, relatively speaking. Internal conditions combined to emit a series of violent eruptions, spewing the iridium that is now found worldwide in a well-defined layer containing concentrations from 10 to 100 times the normal levels of this rare element.

Analyzed sediments from this K – T boundary layer that marks the end of the dinosaur era revealed that the iridium enrichment, along with the other chemical anomalies found there, were deposited over a period of 10,000 to 100,000 years or more.

These anomalies are more consistent with volcanic rather than meteoritic origins. They are significant confirmatory evidence favoring the IN/E facet of the FLINE concept pertaining to eruptions of virgin materials that persistently alter the surface and the atmosphere of Earth, and thereby control the destinies of all its creatures. In this case, the dinosaurs and 90 percent of all genera of protozoan and algae, along with 60 to 75 percent of all species, disappeared from Earth because of the changes wrought via nucleosynthesis within the planet.

During the past few years, a debate has persisted over the source of the iridium in the K–T boundary layer. Some scientists believe that the element came from outer space, brought in by a crashing asteroid that cloaked Earth with a cloud of dust, resulting in darkness, suppression of photosynthesis, the collapse of food chains and ultimately, mass extinction. However, such happenings do not require anything from outer space; Earth is even more capable of stirring up its own mess.

The K – T boundary at Gubbio, Italy was re-sampled in 1990 in detail for both iridium content and magnetostratigraphy by a team that included both the terrestrial and impact proponents. The results confirmed that the iridium anomaly covers about three meters of the vertical section, representing about 500,000 years of deposits. There now seems little doubt that the iridium was not deposited there by the impact of a large asteroid. The evidence clearly shows that the deposits are the result of volcanic actions, thereby adding more powerful support for the FLINE concept.

Many scientists still attribute this demise to an extraterrestrial crash of a huge meteorite or comet. But the evidence against this belief grows stronger. Recently, Nicola Swinburne and her co-workers found glass spherules and high concentrations of iridium within relatively young 61-million-year-old rocks in West Greenland. When researchers detect such evidence in rocks of the K – T boundary age, they often misinterpret it as a sign of an impact.

However, these materials were found in volcanic rock, increasing the probability that an eruption created the spherules and iridium layer. The Greenland rocks also contain large chunks of nickel-iron metal, a principal component of some meteorites. These findings fit precisely into the FLINE concept in which all of these materials are simply ejecta created via nucleosynthesis within Earth's nuclear core.

Ice core drillers at Russia's Vostok Station, atop the great ice sheet of East Antarctica, recently passed 3000 meters – a depth at which the ice is about 300,000 years old. Analyses of air bubbles trapped in the ice have confirmed that levels of carbon dioxide and methane were higher between glacial periods than during them. These findings illustrate how our atmosphere changes over time as a result of subtle changes in Earth's internal conditions that determine the types and quantities of elements (matter) created at any given time within the nuclear core. The creation and expelling of significantly less carbon dioxide and methane to the atmosphere resulted in eras of drastic cooling, now identified as glacial periods (just the opposite of the current warming period).

The most obvious and critical factor determining the rate of planetary evolution is size. The smaller the planet, the more advanced its stage of evolution. Scientists observe Earth and Venus, and classify them as rocky planets (the fourth stage of planetary evolution).

In the fifth and final stage, planets and moons become inactive spheres. Core energies become depleted to the point that no new material can be created within. Outpourings and seismic activity cease. Electromagnetism may or may not be detectable. Mercury, Mars, Pluto and Moon are examples of smaller, nearly inactive spheres. Each has only traces, if any, of seismic activity and electromagnetism remaining.

For example, Moon's very faint seismic activity is revealed in tiny moonquakes, indicating that its nearly depleted core is still capable of creating the weak out-gassings observed in craters. Moon once generated its own magnetic field, which may have been nearly twice as strong as the present-day magnetic field of Earth, according to S.K. Runcorn and colleagues, who used magnetized lava rock from Moon as evidence. This verifies that our Moon's small nuclear core originally was larger that Earth's energy core is today (and both are being depleted).

The proof of the declining strength of Moon's magnetic field is a strong indication that such magnetism is of nuclear energy origin rather than of a steady-state iron core, rock or other origin. This large decline is attributable to the dwindling size of the core as its energy transforms into matter. An iron core, or any other type core, would not dwindle.

From this discussion of the transition phases of planetary evolution, two conclusions can be drawn:

1. All planets and spherical moons evolve through five common stages of evolution via nucleosynthesis by means of natural laws.
2. In any SS, the smaller the mass, the more rapidly it evolves through the five hot-to-cold stages.

One last comment on current beliefs about the origin of our Moon (the giant impact scenario and its most plausible alternative, co-accretion) seems appropriate: In the words of one scientist, "They are more a testament of our ignorance than a statement of scientific knowledge."

Update on Dinosaurs' Extinction

Chemical reaction killed dinosaurs? New clues point to toxic vapor, an AP article appearing in the Atlanta newspapers on December 18, 2000 made the startling announcement: "The rock and dust kicked up by an asteroid impact 65 million years ago was not enough to kill the dinosaurs, according to researchers — but the debris may have sparked a deadly chemical reaction in the atmosphere."

"New studies show the Chicxulub impact crater on the coast of Mexico's Yucatan Peninsula is smaller than once thought, making dinosaur extinction difficult to explain completely. Researchers presented those findings Sunday at the American Geophysical Union's fall meeting." These findings agree with the arguments of the FLINE model during the past two decades; they move scientific progress much closer to its teachings. The article continued: "Since 1980, research on the dinosaurs' disappearance has focused on the 125-mile crater and the [alleged] 10-mile-wide asteroid believed to have created it. Dust from the impact was thought to have blocked sunlight for years." Here, the FLINE model remained steadfast against these assumptions, arguing that the crater is of an ejecta or sinkhole origin rather than via an asteroid impact.

Quoting from the article again: "Now, however, drilling around the Yucatan crater indicates the presence of carbonates and sulfate rocks. The new theory is that these were vaporized by the [alleged] asteroid impact, a process that would have released chemicals that produce sulfur and the greenhouse gas carbon dioxide." The FLINE model goes a big step further by explaining the origins of the carbonate and sulfate rocks, then explaining the clues that substantiate the ejecta origin of the crater.

Quoting again: "The sulfur compounds would have been especially toxic," Sharpton said. "They do nasty things. They form little globules that persist in the atmosphere for some considerable time — decades to a hundred years. They also mix with water in the atmosphere and produce sulfuric acid." However, the major problem here is that dinosaurs died out over a very long period of time, estimated at 10,000 to 100,000 to 500,000 years by a number of researchers. Rather than the short period mentioned in the article, these long extinction times clearly reveal that the demise of the huge animals was the result of gradual changes, wrought by internal nucleosynthesis, in an environment to which the dinosaurs could not adapt sufficiently to survive. These gradual environmental changes via volcanism are accurately explained solely by the two inseparable principles of the FLINE model: Earth's internal nucleosynthesis that drives its evolution through the five common stages of planetary evolution. Once again, we can state with confidence that all planets are self-sustaining entities in which their evolutionary anomalies can be explained without the need to look to outer space for answers.

Quoting again: "How do you initiate the global crisis? It had to be atmospheric chemistry of some sort," Sharpton said. "That's the only way you can transport the effect globally of something that [doesn't] dump the majority of its energy into a single spot on the Earth's surface." In agreement with the FLINE model, Sharpton is correct only in that it had to be atmospheric chemistry of some sort. However, the world-wide atmospheric chemistry changes and transport assumptions in his interpretations that are based on a single impact crater are puny in comparison with the effects of worldwide volcanism.

Certainly, the latter scenario of the well-substantiated FLINE model — e.g., the several separate layers of iridium — can be explained and understood more clearly than with any other known concept.

It is encouraging to think that the article's wide-spread circulation might convince more people of the fallacy of the extinction of dinosaurs via an asteroid impact 65 million years ago, one that simply was not enough to kill the giant animals. After all, they did survive many such impacts during the 140,000,000 years of the Mesozoic era. A better understanding of the actual processes responsible for their demise will be gained by scientists who are willing to investigate the basic principles of the FLINE model's internal nucleosynthesis-evolution connection (formerly known as Internal-Transition-of-Energy-to-Matter).

References: Previous books in this energy series by the author.
NEW CONCEPTS OF ORIGINS: With White Fire Laden, 1986.
THE I-T-E-M CONNECTION: How Planet Earth and Its Systems Were Made, 1990.
The SPACING OF PLANETS: The Solution to a 400-Year Mystery, 1996.

The World of National Geographic's "Volcano", a spectacular video of volcanic eruptions that occur worldwide, was featured on GPTV (December 26, 2000 at 9:00 p.m.). The statement was made that some 1500 volcanoes are active at various times, but the film did not venture into either the source of the voluminous magmatic outpourings or their virgin atomic nature. The illustration of Earth's interior still depicted its core as made of iron rather than of a nuclear energy nature.

While the surprising statement was made that all atoms in human bodies came from within Earth, it did not venture into either the origin of these atoms or how they were created via the processes of nucleosynthesis within the planet's energy core. But the statement does move the FLINE model a step closer to realization and acceptance; the final step will come with the inevitable acknowledgment of the manner in which these human atoms, along with all atoms comprising planetary matter, have taproots deeply embedded in the planet's nuclear energy core. The same principles apply to all planets throughout their five stages of evolution.

The video clearly reveals the status and fallacies of current beliefs about the origin and evolution of Planet Earth (and all other planets).

Four Original Clues to Planetary Nuclear Cores

A number of observations and facts contribute to the conclusion of the 1970s that a nuclear mass, rather than rock (then, the prevalent theory) or iron (the current prevalent belief), exists as Earth's core. First, the three-layered system of energy fuels throughout the crust made it obvious that a central source of nuclear energy was essential for creation of elements and compounds comprising these fuels: gas, petroleum, and coal. The predicted vast stores of deep gas, composed of tremendous volumes of the two elements, carbon and hydrogen, strongly indicated only one possible source for these building blocks comprising methane gas – the starting point for creation of hydrocarbon fuels.

The second and most common observations are the tremendous out-pourings of lava, gases and other matter from volcanoes and rifts throughout Earth's history. Examples include the lava flows that played major roles in building the continents of North America, Europe and India (Deccan Traps), the mile-high multi-layer Grand Canyon and numerous other layered systems throughout the crust.

The third clue is the radioactivity within Earth's crust – known to be a natural byproduct of fission-fusion reactions in nuclear masses. The obvious interpretation of this powerful evidence remains in direct opposition to the persistent belief that credits radioactivity as the <u>source</u> of internal heat, rather than the byproduct of the interior reactions. However, the fact that fission-fusion reactions inside nuclear masses (e.g.; nuclear bombs) do produce radioactivity (as well as heavy elements 99 and 100) positions this as the most logical reason for its presence in the crust. One must question why the radioactivity found in crustal matter is always cold. And how would radioactivity get inside Earth without the presence of a bona fide source to produce it?

The fourth original clue indicative of nuclear cores inside planets and moons can be observed daily: the high mountains and uplifted plateaus found on Earth, Mars, Moon, and more recently, as predicted by this concept, Venus. The presence of such similar characteristics on these spheres is strongly indicative of a common cause: the creation and uplifting of their surface features by very powerful forces within each one. Only nuclear power, in conjunction with isostacy, seemed capable of wielding such magic. These observations and interpretations made it easy to predict that other planets and moons have undergone similar processes. Time and space probes have proven the accuracy of this prediction.

The article *Did Volcanoes Drive Ancient Extinctions?* (*Science*, 18 August 2000, p.1130) presented powerful evidence of the connections between volcanism and extinction events. Jozsef Palfy and Paul Smith "reported that they have improved the dating of both the extinctions and the accompanying large volcanic outpouring. They gathered the latest argon-argon and uranium-lead dates, recalculated them to account for uncertainties peculiar to each, and found a peak in this so-called Karoo-Ferrar volcanism at 183 million years +/- 2 million years."

"The rough coincidence of a large volcanic outpouring and a sizable extinction event brings to five the number of examples of apparent volcanic-extinction correlations including three of the big five mass extinctions. As Olsen pointed out recently (*Science*, 23 April 1999, p. 604), increasingly abundant and reliable radiometric dating techniques have linked in time three of the largest mass extinctions — the Cretaceous-Tertiary 65 million years ago, the Triassic-Jurassic 200 million years ago, and the Permian-Triassic 251 million years ago — with three of the largest flood basalts: the Deccan Traps of India, the Central Atlantic Magmatic Province of northeastern South America, and the Siberian Traps, respectively. And deep-sea extinctions and a turning point in mammal evolution 55 million years ago at the Paleocene-Eocene boundary (*Science*, 19 November 1999, p. 1465) coincide with the massive lavas laid down when Greenland and Europe parted tectonic ways. The coincidences are within a million years or so, as tight as current dating allows." Congratulations to the researchers on their volcanism-extinction dating, a powerful confirmation of the scientific validity of the FLINE model.

"The very big flood basalt provinces are remarkably correlated with extinctions," says Paul Renne of the Berkeley Geochronology Center in California. With these findings, scientists have speculated that episodes combining climatic warmth, massive volcanic eruptions, oceanic anoxia, and bursts of methane may lie behind major extinctions — a giant leap toward vindication of the revolutionary FLINE paradigm. During the past quarter century, the new model has argued persistently and with logical reasons in favor of such volcanism as the basic cause of all extinction events, including that of the dinosaurs. The FLINE model goes a step further by explaining with incontrovertible evidence the fundamental causes of volcanism: internal nucleosynthesis, the force that drives all planets through five common stages of evolution.

All facets of the evolving concept have been presented to various scientific organizations during the past quarter century, but it remains to be examined, vetted and accepted by the greater scientific community. Thus, to preserve it for posterity, a cumulative set of Energy Series publications was initiated in 1975 with *Fuels: A New Theory*. Specific volcanic-extinction connections were explained in *New Concepts of Origins: With White Fire Laden* (1986), then expounded in *The I-T-E-M Connection: How Planet Earth and Its Systems Were Made by Means of Natural Laws* (1991). This was followed by *The Spacing of Planets: The Solution to a 400-Year Mystery* (1996) featuring the revolutionary Fourth Law of Planetary Motion, explaining how the nebulous planetary masses attained their orbital spacing around the Sun. These writings inexorably link the Four Laws of Planetary Motion (FL), internal nucleosynthesis (IN) and evolution (E) via volcanism — three inseparable principles of Nature that apply to all planets. They explain the specific volcanism-extinction connections used in the subject *Science* article.

The FLINE concept makes it easy to understand the fundamental connections between volcanism and extinctions, as well as the basic causes of all other planetary anomalies. It eliminates the need for speculation about asteroids and/or comets being involved in extinctions. There is no need to look to outer space for answers that can be found right here, inside Planet Earth. As scientists dig ever deeper into the FLINE model, their understanding of the new concept will open new avenues to easier understanding of all planetary anomalies.

To: Glenn Strait, Science Editor, *The World & I,* Washington, DC 20005. 05-04-98

Beginning at 6:00 p.m. on May 3, I watched *Dinosaurs: Inside And Out And Then There Were None* on the TV Discovery channel. The film first dealt with the impact of a massive asteroid as the cause of their alleged short-term demise, as described by Bob Barker. This was followed by the new concept of a longer-term extinction via the decreasing oxygen content (from the original 35%) in the atmosphere, which gradually became insufficient to support the huge dinosaurs (as explained by a Dr. Rigby).

As you may recall, this latter version deals with a vital aspect of the FLINE paradigm that goes a big step further by explaining the taproot cause of the decrease in oxygen content: Earth's nuclear energy core that persistently alters the content of our atmosphere and surface features during the ongoing evolvement of Earth through five stages of evolution common to all planets. The section that gives the details can be found in my writings as far back as 1986 (*New Concepts of Origins: With White Fire Laden*, pp. 53-55). Updated versions are detailed in my later books (1991, 1996) and letters written since that time. <u>All the reasons species come and go point directly to the irrefutable fact that evolution and internal nucleosynthesis are inseparable: one cannot exist without the other</u>. Further, all evidence used in support of the theories of Barker, Rigby *et al.* is even stronger support of the new paradigm.

While I am happy to see other scientists contributing supportive facts to the FLINE paradigm, it is ironic that they can get aspects of the concept published so readily via various media outlets while I have struggled for the past 25 years to put these factual aspects together into what now appears to be an irrefutable and flawless paradigm—and cannot get published. I am beginning to doubt the value of copyrights—even though they have been strengthened in recent decades—and to doubt the objectivity of planetary science in general.

After 3½ months, I have not yet received an evaluation of my manuscript, *On the Spacing of Planets: The Proposed Fourth Law of Planetary Motion,* submitted to the Royal Astronomical Society. During that time, I have learned that the Royal Astronomer is Martin Rees, while the Isaac Newton Chair is Stephen Hawking; both are avid supporters of prevailing beliefs about planetary origins. I did write them last week for a status report. The final decision on the proposed Fourth Law and the FLINE paradigm should be very interesting, and should make history one way or the other.

Thanks again for your interest and support.

Recent Clues to Nuclear Cores in Planetary Spheres

A number of clues indicative of nuclear cores have been discovered during the last few years. Foremost among the predicted evidence is the discovery in 1987 that Earth's center is hotter than the surface of the Sun. The inner core has a temperature of about 12,420°F, scientists from the University of California at Berkley and California Institute of Technology reported in the April journal, *Science*. They calculated temperatures of 11,900°F for the boundary between the inner and outer cores, and 8,640°F for the outer core-mantle boundary. In comparison the surface temperature of the Sun is about 10,000°F.

The researchers based their experiments on the assumption of an iron core at Earth's center and a pressure of 49 million pounds per square inch. Their finding was surprising to them because it suggests that the core, not the enveloping mantle, is the source of much [actually all] of the internal heat.

"Thus, the forces that drive the plates and give rise to earthquakes and volcanoes have their origins in the Earth's core," said Thomas Ahrens. "This provides us strong insight into how Earth works."

While this insight is a giant leap forward from the rocky core concept of a few years ago, it falls far short of the true situation inside Earth: a nuclear core with temperatures ranging into the millions of degrees – perhaps into the 100 million degrees range believed to be a requisite for creating uranium, Earth's heaviest stable element.

The core temperatures calculated by Ahrens *et al.* are far too high to have been generated by the slow decay of radioactive elements. Further, such calculations are invalidated by the assumption of an iron core and 49 million pounds pressure, which results in temperatures far below reality.

Viewing these factors in the light of the second law of thermodynamics, one must conclude that the extreme heat was present from the beginning over 5 billion years ago, even before the Earth began forming its crust. This evidence adds dramatically to the FLINE model of an original nuclear mass transforming into matter that formed our atmosphere and crust.

The second important piece of new evidence supporting this concept is the discovery made in the 1980's by seismic tomographers exploring the interior of our planet via sound waves. By slicing open Earth with tomographic techniques and modern computers, geophysicists have uncovered features from the crust down to the core. Topographic maps prepared by scientists at Caltech and Harvard show 'blobs' of hotter material rising from the core-mantle boundary, while cooler masses are sinking from the upper mantle into the interior.

Contradictory to the serene, onion-layer concept, the outer core does not have a smooth, bland surface; rather it consists of alternative deep valleys and mountains. The scale of these deep depressions and elevations is estimated to be between five and ten kilometers – greater than Mount Everest – in some locations. These findings remind one of Sun flares, conforming to the visualized violent images of a nuclear mass encapsulated in the mantle of hot, molten matter of its own making.

The third piece of predicted and confirmatory evidence favoring the new concept came with the discovery in the 1980s of the simultaneous increases in sea levels and polar ice sheets. According to the 1985 National Academy of Science report, sea level is rising about one-tenth inch annually, but scientists don't know why. They have been unable to put the responsibility on the greenhouse effect or on the polar ice sheets.

However, when viewed in the perspective of the FLINE model, the answer becomes obvious: the internal creation of virgin water, as with observable land increases, is an ongoing, never-ending process.

The best examples of virgin water can be seen in the outpourings from deep-sea hydrothermal vents. The Juan de Fuca ridge in the Pacific Ocean has hydrothermal jets spewing jets of mineral-laden water at 662°F in a fairly continuous stream. The minerals are making their surface debut as new matter from Earth's transformation factory.

Scientists believe that mega plumes (large columns of warm water) come from fields of these vents. However, the mega plumes represent an explosion of fluids, like a giant underwater burp. Multiplied many times throughout the World, the result is a steady increase in sea levels. For virgin water and other new matter, there are, of course, many other outlets: volcanoes, rifts, etc.

From as far back as September, 1981, government scientists began analyzing metal-bearing chunks spewed out of an undersea volcano 270 miles west of Oregon. H.E. Clifton, Chief of the USGS's Pacific-Arctic Branch of Marine Geology, reported, "new earth crust is actually being made" by the volcano.

Other strong arguments for creation of virgin materials that form Earth's crust include the "black smokers" discovered in 1979 by a team of American, French, and Mexican investigators. They observed turbulent black clouds of fluid billowing up from chimney-like vents, much like factory smokestacks (which, in reality, they are). The venting fluid, a metal-rich hydrothermal solution was measured at

350°C. Mixing with the ambient seawater causes copper-iron-zinc sulfides to precipitate as fine black particles suspended in the plumes.

The grade of the metals is comparable to that of many ancient massive piles of sulfides on land: 31% zinc, 14% iron, 1% copper, plus small amounts of silver and gold. On the island of Cyprus some 90 large deposits of copper-iron-zinc sulfides occur as saucer-shaped bodies up to hundreds of meters in diameter. They fill depressions in volcanic lavas that erupted on the sea floor some 85 million years ago. Still further back in geological time, similar hydrothermal convection systems were active 2.7 billion years ago in rocks of the Archean period, now exposed in the eastern Canadian shield.

Further verification of an active energy core came on December 5, 2000. Ocean researchers, sponsored by the National Science Foundation, discovered huge, eerie spires of a deep-sea hot water oasis that dwarfed all known sea-floor vents. They are the largest and oldest formations of their kind yet observed in any hot water vent field.

The researchers were stunned by the enormous sizes of the skyscraper-like structures with multiple white-capped towers soaring as high as 18-story buildings, with large diameters of up to 30 feet. The area, at least as large as a football field and submerged 3,200 feet below the surface of the cold North Atlantic Ocean, was appropriately named "Lost City". It actually is the construction work of a system of hydrothermal vents and fissures through which extremely hot water from within Earth's crust is forced upward into the cold sea-water, thereby building up deposits of virgin minerals in the same manner as the younger black smokers described above.

Minerals and lavas, water and land, air, gases, oils, sulfides and other chemicals, etc., constantly pouring from an internal source over eons of time. Such observations continue to add credibility to the theory of Internal Transition of Energy to Matter (I- T- E- M) theory, now identified by more recent terminology as the Internal Nucleosynthesis (IN) concept. In combination with the new Fourth Law of Planetary Motion, it has evolved into the all-encompassing FLINE concept that explains our planetary origins and evolution from the beginning to the present day.

The reason for these continuous outpourings is a crucial point in the concept. The transformation of energy into molecules of matter entails a tremendous expansion, relatively speaking. In such transformation, each molecule expands into far more space than had been occupied by the energy from which it was made. This forced expansion of uncountable molecules is what creates even more pressure and, consequently, ever greater temperature within the nuclear energy core. These dramatic increases in the demand for space, and in temperature, account for the capability of Earth to produce all heavy elements up to uranium.

Simultaneously, the expansions result in a steady, but imperceptible increase in the size of Earth. As it expands, the crust cracks, giving the surface an appearance similar to the cracked shell of a hard-boiled egg.

The fourth piece of predicted and confirmatory evidence for the IN concept can be illustrated by events occurring in Lake Nyos in Cameroon, West Africa. One night in 1986, Nyos experienced a large burst of carbon dioxide from the lake depths. The spreading gas snuffed out the lives of 1700 people. Two years earlier the same type gas had burst from Lake Monoun, killing 17 people. The other 37 lakes in Cameroon posed no immediate danger, although they, too, are nestled in volcanic craters. Scientists remain puzzled by the processes involved and the true source of the gas.

Nothing but a warm, mineral-laden subsurface spring seemed capable of delivering the type of ions identified as increasing quantitatively. The composition and warm temperatures of the bottom water point to a hot, deep spring feeding it from below. Much like deep-sea hydrothermal vents and our familiar volcanoes, the volcanic craters holding the lakes simply serve as outlets for the virgin gases. Worldwide and on other planets and moons, many similar outlets exist for such out-gassings.

Methane and hydrogen sulfide are examples of gases discovered on a number of SS bodies besides Earth. Hydrogen sulfide, found for the first time (1989) outside our planet, is present on the surface and in the atmosphere of Io, one of Jupiter's moons. More recently and seemingly stranger, French astronomers reported the presence of hydrogen sulfide in the comets Austin and Levy. These crucial discoveries confirm that gases are consistently present in all SS bodies suspected of being powered by nuclear interiors. Comets, as we shall see later, are no exception.

A good example to consider here is the gaseous atmosphere of our nearest neighbor, Venus. Known for some time, its bright envelope is 96% carbon dioxide with a substantial admixture of argon. Such an atmosphere argues strongly for out-gassing from a hot interior capable of selectively creating such voluminous gases. Further, recent mappings of the surface of Venus reveal that its unique surface characteristics, conforming to predictions, are readily explainable by principles of the FLINE model discussed previously. Its excessive atmospheric heat contributes to its uniquely flowing surface features.

More Evidence Favoring Earth's Energy Core

An article on *The Globe Inside Our Planet* (*Science News*, Jul 25, 1998) states that Earth's inner core consists of solid iron. In the accompanying article, *Solid Core Proof Ends Half-Century Search*, researchers have observed shear waves passing through the inner core, which they interpreted as evidence of a solid inner core. To quote one statement: "Seismologists have not doubted that the inner core is solid because several lines of indirect evidence point toward that conclusion." This conclusion of the two articles raises some key questions: Aren't the electromagnetic fields of Earth and our Sun and stars all produced in identical manner? If Earth's electromagnetism is created via an iron core, does this mean that our Sun's powerful electromagnetism that stretches for billions of miles also is produced by an iron core inside this huge sphere? In view of the very short ranges of magnetism created by mankind's iron-core magnets, how can electromagnetism generated via an iron core remain effective over billions of miles of space? Do all stars possess iron cores that generate their powerful electromagnetism?

Serious consideration of these questions point to a more powerful source of electromagnetism residing inside every active sphere. Rather than a solid iron core producing magnetic fields, a much more powerful source is inherent in the core of every active sphere. That source is nuclear energy. *Without this source of energy to drive evolution via the processes of internal nucleosynthesis (IN), in which the 90 atomic elements of Earth's atmosphere, land and water are created, evolution (E) of our planet (and all planets, moons, Sun and stars) would not be possible. In any active sphere, IN and E are inseparable; one cannot exist without the other. When that source of energy is depleted to the point of becoming a solid core, evolution ceases and the planet or moon in question enters its fifth and final stage of evolution: a dead (inactive) sphere with little, if any, electromagnetism.* A plethora of supportive evidence is detailed in *THE SPACING OF PLANETS: The Solution to a 400-Year Mystery* by A. Scarborough (1996).

The final piece of the enigmatic puzzle of how our Solar System came into being is the recent solution to the proposed Fourth Law of Planetary Motion detailing how the planets—while still in their original Sun-like energy stage—attained their orbital spacing around the Sun. Thus, the planets began their evolvement through the five stages of evolution common to all planetary spheres—from hot to cold stages. We have only to observe the Sun, Jupiter, Uranus, Earth and Mercury to see examples of these five stages of evolution—all in full accord with size and with all natural laws.

The strong supportive evidence furnished by the Four Laws of Planetary Motion (FL) that properly spaced the energy masses around the Sun, together with the subsequent close relationship of IN and E, provide a valid, testable and definitive scientific concept of planetary origins and evolution via natural laws: The FLINE paradigm.

One remaining question: Until the results of shear waves passing through a dense energy core are known, how can researchers be certain that the results of shear waves passing through Earth's inner core can be interpreted as clues to an iron core rather than to a dense energy core? A comparison of such results, if ever obtained, would add the fourteenth clue to the FLINE paradigm's evidence of the true nature of Earth's core: Sun-like energy—just as Descartes first predicted.

Corroborative Evidence for Nucleosynthesis Within Planets and Moons. (1)

One of the best illustrations of how planetary matter came into being via the processes of internal nucleosynthesis and subsequent polymerization is the creation of hydrocarbon fuels (i.e., gas, petroleum, coal), erroneously called fossil fuels. In the beginning, the countless numbers of atoms of carbon and hydrogen were (and still are) created via nucleosynthesis in Earth's energy core. One carbon and four hydrogen atoms join readily together to make one molecule of methane gas. When two of these molecules link together via the process known as polymerization, ethane is produced. Further polymerization produces propane gas composed of three methane molecules, followed by butane gas consisting of four methane molecules.

When five molecules of methane combine, they form pentane, the lightest weight of the oils found in lightweight petroleum. As a variety of other elements become involved in the mix, the polymerization and cross-linking continues onward into the crude petroleum stage of evolution. Scientists have found that every small sample of crude petroleum contains tiny particles of coal — conclusive evidence of the ongoing evolution of petroleum into coal.

When huge volumes of petroleum were forced onto Earth's surface, they inundated the low-lying swamplands and encapsulated the plants and animals in a sea of oil that began cooling, polymerizing and cross-linking into thick beds of coal. In the 1830s, a scientist named William Logan discovered the imprints of these live plants in coal. From this and similar evidence, Logan initiated the fossil fuels theory (FFT) by concluding that coal was made from plants. However, he overlooked the fact that such imprints of live plants can be preserved only through sudden encapsulation — in this case, by the liquid petroleum that solidified into coal. Those imprints of live plants in coal, including tree trunks, had been preserved only by their sudden encapsulation by the gushing petroleum. The tree trunks became carbonized masses, in a manner analogous to that of petrified trees.

According to this Energy Fuels Theory (EFT) of 1973, these hydrocarbon fuels should exist in an irregular pattern of a three-layered system throughout Earth's crust — and they do (Fig. 6). Coal is found on or very near the surface, petroleum is found at moderate depths, while the vast majority of the huge volumes of gas exists at deeper levels. Of course, noticeable overlapping of the three layers do occur as the gas and oil are forced upward through paths of least resistance — but not enough to discount the fact of a three-layered system of fuels. The varying mixtures of gas and petroleum at moderate levels, and the fact that every lump of coal contains traces of methane and petroleum adds corroborating evidence to this EFT, Nature's way of making these hydrocarbons.

The huge volumes of methane hydrates discovered throughout Earth's crust during the 1980s were predicted in the 1970s, and are explained, by this EFT. And at the peak of the energy crisis in the mid-to-late 1970s, the knowledge that petroleum was not made from fossils allowed me to predict confidently in print the current price of around $1.00 per gallon of gasoline and its bountiful supply of the 1990s (*Undermining the Energy Crisis*, 1977, 1979). You may recall the news headlines of near-panic fears of the 1970s of running out of these fuels by 1990, while the price of gasoline would soar above $5.00 per gallon.

Much more powerful corroborating evidence for the EFT can be found in the scientific literature while the powerful evidence against the FFT continues to mount. The EFT is a typical example of how all planetary matter was, and is, created via internal nucleosynthesis, polymerization and cross-linking — the natural processes of evolution.

The processes used by Nature to create all matter are reversible by mankind. For example, coal can be converted back into oil, which can be changed into molecules of gas from which the atoms can be isolated and then split to release the original powerful energy with which they were created.

Earth Story - A TV Series

On September 20, 1998, the *Earth Story* series was initiated on TV Station TLC. The first two parts of the program's six segments were shown from 9:00 to 11:00 pm, EST. The first part of the video was a brief discussion of some important geological discoveries (Wegener, Hutton, Hall, Vine, Sykes, etc.) in chronological order. Conspicuously absent was the concept of isostacy first recognized by Aristotle and later identified and expounded by Dutton.

Unfortunately, the entire TV series apparently will be based on the erroneous Big Bang (BB) hypothesis in which the overwhelming evidence against this explosive concept far outweighs the speculative beliefs and misinterpretations of data offered in its support. In turn, through no fault of their own, advocates of the BB are forced into misinterpretations of their findings about the planets of our Solar System (SS) — which, in every case, can be interpreted definitively via the FLINE paradigm of the origins, orbital spacing and evolution of planets.

The formation of planets via the condensation/accretion hypothesis of the origin of our SS is an excellent example of the situation in which the Four Laws of Planetary Motion and the principles of evolution are blithely ignored. This is borne out by the fact that planetary evolution is not possible without internal nucleosynthesis (IN), the engine that is absolutely indispensable for driving planets through their five stages of evolution. The vast amount of supportive evidence, in which no relevant discovery can remain outside its realm, assures that this basic principle—the First Principle of Evolution (FPE)—is both indispensable and indisputable.

According to the FPE, planetary evolution would not be possible if Earth's center consists of either a rocky core (1980s) or an iron core (1990s). The existence of any type of core other than a nuclear energy mass in any planet is possible only in inactive (dead) spheres in which evolution has reached the end of its cycle. Certainly, the prevailing idea of an iron core at Earth's center, as emphasized in the TV series, needs to be reexamined if we are to truly understand planetary evolution.

Wegener's concept of continental drift contains much verifiable truth—as far as it goes. However, it needs to explain that the days of continental drifting, per se, have ended; the continents have been locked into their positions by Earth's ever-thickening crust, while the planet imperceptibly expands in size. Although it still prevails over the expansion concept of the FLINE paradigm, the theory of subduction and recycling of the crust is an idea whose time has come and gone. The continual creation of virgin matter via IN accounts for this ongoing expansion, which, in turn, accounts for the illusion of continental drift and for the ubiquitous cracking of Earth's crust—and its "plates"—and is a key factor in earthquakes and in the cause of Dutton's isostacy.

While *Earth Story* is interesting and helpful, it does leave wide gaps in the understanding of planetary origins and evolution. A series based on the FLINE paradigm, underpinned by the Little Bangs Theory (LBT) of the origin of the Universe, could fill all of these gaps.

Ringing Earth's Bell: What makes our planet constantly quiver? (*Science News*, 7-4-98)

In mid 1998, two teams of Japanese researchers reported discovery of the faintest of reverberations, an inaudible incessant pulse coming from Earth itself. "This planetary ringing has set seismologists around the world searching for an explanation... Unlike the bursts of vibrations from earthquakes, the mystery vibrations appear to resonate continuously... In the past, seismologists had thought that these slow planetary stirrings came only with energetic quakes. But researchers are now starting to tune into extremely feeble oscillations that ring all the time." They have reached a provocative conclusion: "The observed 'background'-free oscillations represent some unknown dynamic process of Earth."

Logic, tests and calculations have eliminated all suspected causes (earthquakes, wind, tectonics, etc.) of the ringing. Quoting the article again, "At this early point in the investigation, however, researchers must also consider geological forces as a source of the vibrations. 'It would be exciting if these observations lead to the discovery of some slow, deep process,'" says Kanamori.

These observations will indeed lead to the 'discovery' of a slow, deep process that is the heart of creation of all planets: a nuclear energy core. Even more exciting is that this 'discovery', when added to the 13 previous (1973-1995) clues to the nuclear nature of Planet Earth's core, should be sufficient to convince skeptics of the scientific validity of the FLINE paradigm of planetary origins. One of those clues is the recent discovery that our Sun also rings like a bell; from this fact, we can conclude that all other stars and nuclear masses also ring like a bell.

An exciting aspect of this turn of events is the probability that all planetary spheres of the Universe pulsate in accord with their stage of evolution—until they enter the fifth and final stage of evolution when their cores are depleted. Thus, such observations can serve as a tool to determine whether the core of any planetary sphere is active or depleted.

The results and conclusions raise some interesting questions. Was the project inspired by previous knowledge of the already-proven nature of Earth's nuclear energy core? Were the researchers already aware of the "unknown dynamic process of Earth," and, if so, did they design the project in order to be able to reach their "provocative conclusion"?

In any case, their results and conclusions add corroborative evidence of the nuclear energy nature of the dynamic core of Earth (and all active planets): the basic foundation of my evolution-via-internal-nucleosynthesis (IN/E) concept of 1973-1977 that inevitably led to the Fourth Law of Planetary Motion (FL) to become the FLINE paradigm (1995) of planetary origins and evolution.

Fusion Energy and Magnetism

Two articles in *Science* (6 August 1999) are interesting and timely: *Common Ground for Fusion* by James Glanz (p820-821) and *Fusion Power From a Floating Magnet* by James Riordon (p821-823). The move away from the doughnut-shape of the tokamak and into more spherical shapes (e.g., Spheromaks) is a giant step in the right direction for fusion reactors. To quote: *"The Levitated Dipole Experiment (LDX) is a 5-year study of a plasma confinement scheme inspired by observations of ionized gases trapped in the magnetic fields of planets like Jupiter and Earth,"* [the Sun and other planets]. Again, from the article: *"Levitated dipole reactors, in contrast, are the least complex fusion machines yet conceived. Current-carrying loops (like the super-conducting ring at the heart of LDX) and common bar magnets generate dipole fields, the simplest* of magnetic field configurations. So do planets, such as Jupiter. It was the Voyager II spacecraft's detection of plasma trapped in the fields of Jupiter's magnetosphere in the late 1980s that inspired Akira Hasegawa...to propose the dipole design for the fusion machine."* (Note: *Nature always prefers the simplest way; mankind must copy.)

During the past two decades, I have written to fusion researchers that to understand and accomplish fusion on a commercial scale, they must look to the planets and duplicate the manner in which their fusion is powered by nuclear energy cores. In looking to the planets, Hasegawa and I apparently followed the same line of reasoning, but from different perspectives; mine came from the evolving FLINE model in the late 1970s, and was further verified by the Voyager II findings in the 1980s. *The key point: the powerful support of the ideas derived from planets, and now being incorporated into the new fusion machines, do offer additional scientific validation of the FLINE model of planetary origins and evolution.* Whether or not the researchers are aware of this vital key point remains unknown.

It is indeed fortunate that Hasegawa and others were in positions to get researchers started in the right direction on the production of fusion energy via the same processes utilized in planetary cores. The understanding of either of the fusion processes (in planets or in machines) will lead eventually to understanding the other, along with full validation of the FLINE model. Solutions to all anomalies of the SS should follow easily and soon after this validation. But one must question why it is necessary to wait for further validation of the new concept when the mysteries of solar system origins, planetary quakes, lightning, volcanic eruptions, the origins of moons, etc., can be solved now simply by assuming the scientific validity of the FLINE model.

Why Electromagnetic Field Strengths Vary (1994)

One of the most puzzling anomalies in science concerns the origins of the electromagnetic fields embodied in the spheres of our SS. Why do they exist and why do their strengths range so widely between very weak and very strong from sphere so sphere? Do they remain constant for billions of years?

A comparison of the field strengths of Venus, Earth, and Mars should offer some insight into the reason for the wide range of electromagnetic strength from sphere to sphere. Venus is almost as large as Earth, but Mars is only one-seventh as large. The fields of Venus, a very slowly rotating planet, and of Mars, a very small planet, are much weaker than Earth's. At this writing, NASA scientists have concluded that if Mars does have an intrinsic magnetic field, it is not of any consequence. Likewise, although nearly the size of Earth, Venus has been identified as a planet with little, if any, electromagnetism. Additionally, our Moon, being small and having only one rotation for each revolution around Earth, shows no detectable magnetic field.

Further, the highly tilted and offset magnetic field of Uranus, a midsize sphere, is the strangest one among planets. The offset from the planet's center causes its magnetic field strengths at the surface to vary by a factor of ten between the north and south magnetic poles. Adding to the evidence, the two largest planets, Jupiter and Saturn, both with faster rotational speeds than Earth, can be expected to have much stronger EM fields. And they do. Mercury, the smallest planet, has a field strength of only one percent of Earth's, and a very thin atmosphere. Both clues are indicative of the presence of a small, active remnant of a nuclear core.

There are two basic principles of electromagnetism: Electricity in motion produces a magnetic field, and a magnetic field in motion across an electrical field produces an electromotive force. Combining these principles with the above observations, one can make a general rule: The electromagnetic field strength of each SS sphere is a function of its core size, speed of rotation, (and perhaps its velocity of revolution) and the angle of inclination of the axis.

The electromagnetic field strength (EFS) rule can be true only in the concept of nuclear cores. Thus, a fifth new confirmation of this revolutionary IN idea appears to have been established. If, and when, proven, it will further verify the validity of the total FLINE concept.

In contrast to the new concept, prevailing beliefs dictate the existence of various types of cores in individual spheres, which serve as generators of their magnetic fields. For example, Earth's core consists of iron, either in the molten state or a crystallized solid. Using Earth as an example, one of the main problems with such non-nuclear cores can be seen in a comparison of potential strengths between a field created by an iron core versus a field created by a nuclear energy core; e.g., the Sun. The powerful, extensive magnetic field created by the Sun's nuclear mass reaches many billions of miles to the edge of the SS – billions of times greater than the size of the Sun.

Experiments with iron magnets and electric generators quickly reveal their very limited ranges: only relatively small multiples of their sizes. In this perspective, the choice of a nuclear core rather than an iron or a generator type appears more logical. Thus, the fifth piece of predicted and confirmatory evidence (the ninth clue) favoring this revolutionary idea of a nuclear core in each of the bodies comprising our SS becomes more firmly established.

The Mysteries Of Earthquakes

The abstract (1992 Western Pacific Geophysics Meeting), *Infraplate Earthquakes and Crustal Horizontal Temperature Differences in Europe,* by I. Stegena told of three series of tests that correlated the numbers of earthquakes in specific time frames with the geothermal temperature gradients in those areas.

In the Pannonian basin, temperature differences in 1 km depth are compared to earthquakes that occurred between 1859 and 1958. It was found that 95% of the earthquake energy in that period occurred on that half of the basin where horizontal geothermal gradients are large.

In the West European area, 84 of the 93 quakes between 1901-1955 occurred in that third of the area where the horizontal gradients of heat flow density are the most abrupt.

A third study carried out in East Europe gave a similar result in which 81% of the total quake energy between 1901-1973 burst out on only 20% of the area (where the horizontal changes of heat flow density are most abrupt).

The concordant results of these investigations, in which the epicenters are lying mostly on places of large horizontal temperature differences, suggests that the sporadic infraplate quakes are generated by thermal stresses and relaxation. The significant conclusion reached by Stegena is that there is no expressed correlation between tectonics and epicenters: the earthquakes of the area are not tectonic quakes sensu stricto.

At the same meeting in 1992, the abstract by Mary Ann Glennon, on the subject of deep-focus earthquakes beneath the island of Sakhalin, reached the conclusion that at least part of the observed pattern residuals is due to path effect away from the source, not that of subducted slab.

The so-called 'rim of fire' virtually surrounding the Pacific Ocean conveys the notion to many minds that earthquakes and volcanic actions should occur only along the boundary lines of two or more adjourning tectonic plates. However, there are too many exceptions to this assignment to make a general rule. For example, during the week ending July 7, 1995, a *Chronicle Feature* map pinpointed 14 significant earthquakes that had occurred during the previous seven days. Of this number, ten were inland, well away from known tectonic plate boundaries. Five of the ten quakes occurred within the USA.

In October, 1992, Georgia Tech scientists announced that new analyses of molten, pressurized rock, minerals and other materials making up the mantle deep in Earth may provide insights into deep quakes, volcanoes and even the formation of the planet.

Scientists studying the mantle, which extends from about 60 miles to 1,800 miles below the surface, say the region consists of two distinct layers containing different properties of key minerals. The difference between the two layers could help account for certain deep quakes and volcanoes that cannot be explained by conventional theories.

As far as they go, these findings and those of Stegena and Glennon are in complete agreement with the FLINE concept. They, too, indicate the necessity of taking another look at current beliefs about the nature of earthquakes. The next step should be to question the source of the molten magna and its specific role in quakes. The answers will point the way to a better understanding of the nature and relationship of quakes and tectonic plates.

Most scientists now interpret their findings about quakes strictly in the perspective of plate tectonics in which rock slippage at plate boundaries cause the earthquakes. In turn, these beliefs are linked to the Accretion Disk hypothesis, a supportive branch of the Big Bang theory.

However, when viewed in the perspective of the FLINE model, each booming epicenter of every quake pinpoints a powerful explosion that rocks the magma that causes the vicious or gentle shakings, cracking, sinking and uplifting of the crust. These events and the cracked eggshell-type surface they create worldwide attest to the powerful forces constantly at work deep within Planet Earth. This compelling new cause-and-effect version offers genuine opportunities for understanding the true nature of earthquakes - and all other planetary quakes.

Although not consistent in their warning signals, earthquakes do emit pre-quake clues: sudden changes in emissions of gases, electrical signals, and swelling of the ground under tremendous pressures. All of these signals can be traced back to the nucleosynthesis processes in Earth's nuclear energy core.

The Warning Of Precursory Signals

The evidence for precursory warning signals continues to mount. An amateur who studies radio signals had warned that an earthquake capable of devastating damage would hit California within three days. Jack Coles, a former stereo salesman with no college degree, operates the Early Warning Earthquake detection network out of his home in San Jose. The fax he issued on a Saturday in January, 1994 said he had received reports of "increased radio signals, magnetic anomalies and many cases of electrical problems." He warned that the results could mean an earthquake measuring more than 6 on the Richter scale.

The quake that hit two days later in Southern California measured 6.6. Reporters had ignored Saturday's message.

Coles had started developing his theories a few years back when a radio he was repairing started making strange noises. "About four hours later, we had a quake, and I wondered if there was a connection," he said.

Scientists at the USGS in Menlo Park find little to back Coles' theories. "We sent three scientists over there some time ago to look at his stuff. We couldn't make any sense out of it," stated one scientist, (an iron core advocate).

A study published in September, 1992 suggests that the eruption patterns of an 'Old Faithful'-type geyser (near Calistoga) in California could give warning of impending large quakes. In the report in the journal *Science*, scientists said they entered into a computer the record of the geyser eruptions since 1971, and then compared the eruption pattern for months around major quake events.

They found that three major quakes within 155 miles of the geyser occurred within one to three days after the geyser's eruption pattern underwent abrupt and dramatic changes. A mathematical examination of the Calistoga record eliminated both chance and the effects of rainfall as an explanation for the abrupt pattern changes preceding the quakes.

The Kobe Earthquake Signals

In January, 1995, as aftershocks of the destructive Kobe earthquake continued to rumble beneath the region, reports of aurora-like flashes just before and after the deadly tremor were announced by a Japanese professor. Tamenari Tsukuda of Tokyo University said one of the most intriguing sights was a flash that streaked from east to west about eight feet above the ground shortly after the quake.

Phenomena like this are believed to be due to electrical and magnetic waves "generated by the grinding of Earth's crust", when interpreted in the "rock slippage" version of the cause of quakes. In the IN version, such phenomena are generated before, during and after the huge explosion at the quake's epicenter, while rock movements are the results of the powerful blasts.

Precursory signals that forewarn of pending quakes are very real, although not always consistent and not yet dependable. All signals can, and some day will, be traced to their source: Earth's nuclear core.

Radon concentration in ground water increased for several months before the Kobe earthquake. From late October, 1994, the beginning of the observation, to the end of December 1994, radon concentration increased about fourfold. On January 8, nine days before the quake, the radon concentration reached a peak of more than ten times that at the beginning of the observation, before starting to decrease. These radon changes apparently were precursory phenomena of the disastrous quake.

Chloride and sulfate ion concentrations of ground water issuing from two wells located near the epicenter of the Kobe quake fluctuated before the magnitude 7.2 events on 17 January 1995. The samples measured were pumped groundwater packed in bottles and distributed in the domestic market as drinking water from 1993 to April, 1995. Analytical results demonstrated that the concentration of both ions increased steadily from August, 1994 to just before the quake. Water sampled after the quake had much higher ion concentrations. The precursory changes in chemical compositions apparently reflect the preparation stage of a large earthquake.

These precursory changes in concentrations of radon and other chemicals do not appear to have any connection with strain buildup in rocks. Rather, they appear to lead in a direct path to internal nucleosynthesis as the source of the changes that account for the basic cause of earthquakes.

Depth as a Safety Factor

On June 8, 1994, a powerful earthquake sent panicked Bolivians, Chileans and Brazilians into the streets, and was felt as far away as Canada. Power outages occurred in parts of Chile and Bolivia. Although it was perhaps the biggest deep-focus quake of the century, there were no reports of major damage or casualties.

Unusually deep at 400 miles beneath Earth's surface where solid rock does not exist, the widely felt quake caused very little damage. The lack of rocks there allowed the shock waves to travel through the magma and far into the distance, dissipating the explosive energy quickly and safely.

Scientists have learned that our Sun has radical pulsations that make the solar surface contract and expand like a ringing bell. In addition, like a bell, Earth has its own natural frequencies – or normal modes - which start ringing if the globe is hit hard enough – as was the case here. The most persistent of these modes causes the planet to expand and contrast every 20 minutes, almost as if it were breathing. This mode can be detected even three months after a great quake. "This is really going to change the level of our information about the deep Earth," one seismologist said.

Contrary to this magnitude 8.2 quake, the much smaller, shallower, magnitude 6.7 quake in Northridge, California killed 61 people and caused at least $20 billion in damage.

Scientists have yet to unravel exactly the cause of deep earthquakes. According to one prominent, but speculative theory, deep tremors occur when increasing pressures cause minerals in the ocean crust to undergo a sudden structural transformation!

Such speculation is not necessary; the cause is obvious. In any size quake, the deeper the explosion's epicenter, the less the danger at Earth's surface. This holds especially true when deep explosions occur where no rocks exist. If no rock exists there, then rock slippage cannot be the cause of the quake. If not the cause, crustal rock movement must be the result of a powerful explosion at the epicenter of each quake, no matter how deep or shallow. If below the level of solid rock, Earth's bell will ring until the shaken magma settles down again.

Why The Explosions?

In April, 1992, an explosion packing the power of an earthquake ripped open an underground propane gas pipeline, killing one person, flattening nearby mobile homes and shaking buildings more than 140 miles away. It registered 3.5 to 4.0 on the Richter scale – as strong as an earthquake that could cause slight to moderate damage.

"It was just a big bang, a tremendous bang," stated one survivor. One child was killed, and at least 18 persons were injured, three critically.

The explosion occurred in a rural area seven miles south of Brenham, Texas, a community of 12,000 about 70 miles northwest of Houston. Officials suspect gas at a liquid petroleum storage and pumping facility collected in a ravine and was ignited by a car or a pilot light in a home.

A most interesting point here is the comparison of this powerful explosion to a 3.5 to 4.0 quake. One must wonder how different or how much alike the two events actually are.

The Tiny Mystery Of Polonium Halos:
Creationism, Big Bang or the FLINE Model?

In the book, *Creation's Tiny Mystery*, Robert V. Gentry presents a good argument for Creationism, using polonium halos found in granite as evidence supporting this belief in opposition to the evolutionary viewpoint of science. The book was sent to me in June, 1995 by Glenn C. Strait, Natural Science Editor of *The World & I* magazine. His cover letter asked how the presence of these tiny halos in granite could be explained by the FLINE concept.

Under the microscope, these halos show a tiny radioactive particle at the center of concentric ring patterns in the granite, much like the bull's eye at the center of the rings. Because of their radioactive origin and their halo-like appearance, these microscopic ring patterns became known as radioactive halos.

After reading the interesting book, the essence of my reply can be summed up in two short paragraphs: While I do not agree with the conclusion that his discoveries of polonium halos in granite support Creationism, per se, I am excited about two aspects of these findings. First, they offer strong support for the evolution of planets via internal nucleosynthesis (IN), which, in turn is powerful evidence against the Accretion Disk hypothesis of the origin of the SS (a vital factor in the Big Bang theory).

Second, the short half-life of radioactive polonium gives solid assurance that they could not, by any stretch of the imagination, have been formed in distant supernova explosions (as claimed in the Big Bang Theory) and survived the eons that supposedly elapsed before they became a part of Earth's crust. Thus, in the prevailing beliefs about planetary origins, it is impossible for polonium to be a primordial constituent of Earth's granite.

The evidence clearly shows that the halos had to have been formed in situ – and they were. Only in the FLINE model would this be possible! Both the origins of Precambrian granite and the polonium halos therein can be explained readily in the new perspective – a concept firmly structured on the solid foundation of the SS's geometric origin that proved crucial in deriving the Fourth Law of Planetary Motion on the spacing of planets.

Granite is the foundation rock of Earth's ever-thickening crust. It was among the first layers to solidify atop the molten mantle worldwide to form tenuous sections, and eventually a permanent foundation on which layers gradually built – and are still building. How did the compositional elements and compounds form into granite with polonium halos inside?

Polonium is a radioactive metallic element belonging to the uranium decay series. It occurs naturally in pitchblende as a decay product of radium, and can be produced artificially by bombarding bismuth with neutrons. Its most stable isotope has a mass number of 210. Polonium, with a half-life of about 138 days, decays into an isotope of lead by giving off alpha rays. One polonium isotope is the product of the radioactive decay of radon, a common gas that still emanates continually from deep within Earth.

The short half-life of polonium-218 of three minutes means that every three minutes, one-half of its remaining mass will decay. If created along with other elements within Planet Earth, it would be no surprise that traces of any and all polonium isotopes are found in granite. The isotopes formed in situ during cool-down of the granite mix containing the polonium and other elements.

The textures and composition of granite give important clues to its formation processes. Scientists know that these foundation rocks have coarse-grained, crystalline textures, which are only found in rocks that cool slowly from a hot molten mass. The process can be observed by crystallizing compounds in the laboratory; these experiments always form crystalline textures.

These clues point to the only apparent explanation: the elements and compounds comprising Earth's hot mantle-like, thick liquid surface some 4.6 billion years ago came together and slowly cooled to form granite. Radioactivity was at its peak surface performance; huge numbers of large particles of radon, polonium, etc., continually bombarded the coagulating granite; particle entrapment was common, followed by decay that left their halo marks in the granite.

Through the eons the newly-mixed materials from the hot mantle continuously contributed to the ever-thickening crust. The process continues today, and will do so at diminishing rates until all the energy of Earth's core is expended in the processes known as nucleosynthesis.

Thus, to the previous nine clues to Earth's nuclear core, we can add one more: the tiny mystery of polonium halos in granite. While adding much support for the FLINE concept, these tiny clues to Earth's origin clearly aim their arrows at both the hearts of Creationism and the Big Bang Theory.

Galileo's Stunning Probe Into Jupiter

A stunning weather report from the probe launched into Jupiter's atmosphere (7 Dec 1995) by the Galileo satellite hit the news media. The report revealed that the skies were hotter, windier, drier, and clearer than forecasters predicted. In fact, the preliminary data from the probe is so shockingly different than expected that it inevitably will lead planetary scientists to rethink not just the meteorology of the gaseous planet, but its very origins. According to the understatement of one Galileo project scientist, the data "doesn't fit very well. In fact, it's darn uncomfortable."

Contrary to this situation, the report's results are very exciting in that they are precisely what one would expect when interpreting the data in the perspective of the FLINE model. The best way to verify this statement is by comparing the interpretations of the data in the two different perspectives: the current view of Jupiter's origin versus the FLINE view.

The temperature at Jupiter ranged from -144°C at the top of the ammonia-cloud covered atmosphere to +152°C (306°F) at only 600 km (360 miles) into the thick blanket of clouds covering the giant planet whose diameter measures 142,980 km (86,000 miles). Pressures ranged from 400 millibars to 22 bars over the same descent path, compressing the gases to densities up to 100 times greater than previously postulated. The extremes of temperature and pressure created a vertical convective motion in the atmosphere, stirring up turbulent winds more than 50 percent stronger than predicted. These winds were constant throughout the probe's descent.

In theory, the probe should have passed through a region where wind speeds drop to zero. However, it never reached such a point. Contrary-wise, the wind speed increased with depth, which led several investigators to speculate that the energy source driving the circulation of the Jovian atmosphere is probably escaping from the interior. The answer moves us a giant step closer to truth.

In the perspective of the FLINE concept, one would predict conclusively that the engine driving Jupiter's powerful winds is the internal heat source: its nuclear energy core. The core also accounts for the thick, extremely dense blanket of gaseous clouds and the high temperature at the depth of only 600 km. Jupiter's core temperature, even though well-insulated, was recorded by Pioneer 10 in the 1970s at 30,000°C, but this will prove to be a very conservative figure when more accurate measurements can be made.

The probe's helium abundance detector recorded that the outermost regions of Jupiter now contain much less helium than the planet started with, a figure calculated from the helium-to-hydrogen ratio in the Sun. Current beliefs go with the suggestions "that Jupiter's helium is now condensing into droplets under the deep interior's megabar pressures; the droplets then fall even deeper into the planet. So, the gravitational energy released as heat by the fall of the helium raindrops must, in fact, be fueling Jupiter's infrared glow, which is brighter than anything the solar energy reaching the planet could account for."

A more sensible explanation is that Jupiter's core, in strict compliance with natural laws of evolution from energy to matter, simply is producing a smaller percentage of helium as it gradually evolves into the stage that produces less helium and/or more heavier elements. Rather than being produced by helium raindrops, the heat fueling Jupiter's infrared glow emanates directly from the nuclear core, just as predicted by the FLINE model.

Further, the probe revealed Jupiter to be much drier than anticipated and relatively free of condensation. The observation that its atmosphere contains water concentrations equal to that of our Sun left investigators wondering. However, both the helium and water concentration are in line with predictions of the FLINE concept in that they reveal the close relationship of the two masses. They are clear indications that Jupiter has evolved into the second stage of evolution from the energy stage by covering itself with the initial blanket of chemicals that still match the Sun's production. As the huge planet evolves toward the rocky stage, its production of chemicals will gradually lean more and more toward the heavier elements, and its higher-than-expected radiation doses will gradually decline as more and more of its core energy is transformed into encapsulating matter.

Scientists can learn a great deal about planetary evolution through the realization that planets actually do evolve from energy masses through five common stages. Jupiter is an excellent specimen to study for details of evolutionary changes as the giant planet progresses through the second stage.

As for water clouds, mere wisps of water particles were found in the atmosphere. Even many of the heavier elements – carbon, oxygen, and sulfur – as well as neon were found to be at lower-than-expected concentrations. Here again, these findings are revealing great details about planetary evolution. There simply is no need to use the excuse that Galileo's probe dropped into a very rare "hot spot" to explain these undoubtedly stunning results. With winds at 531 km/hr, temperatures would tend to level out any such spots. It is a safe bet that the results are truly representative of the atmosphere on Jupiter.

Just as Galileo himself drove the dagger into the heart of the Ptolemaic version by promoting the Copernican idea of a Sun-centered Solar System, the Galileo orbiter has plunged a dagger into the heart of the prevailing dogma about planetary origins and evolution.

THE FLINE PARADIGM: Creation By Means Of Natural Laws.

Abstract

All efforts since the initial attempts of Johannes Kepler in 1595 to solve the mystery of how the planets attained their orbital positions around our sun have failed to produce a definitive solution to this enigma. To understand the mystery of the origins of solar systems and the evolution of planets, one must first explain, beyond doubt, the critical relationship between the current spacing of our planets and the forces that power the evolution of these celestial spheres. Prevailing concepts fail to explain this inseparable connection.

The key to understanding this relationship resides in Kepler's First Three Laws of Planetary Motion and the solution to the proposed Fourth Law of Planetary Motion. Together, these Four Laws offer a valid explanation of the dynamic manner in which our Solar System came to be, while providing a solid foundation for understanding the forces that drive the evolution of planets. The complete mathematical solution is detailed.

A valid solution must provide a solid foundation for supplying definitive answers to all anomalies of our solar system; e.g., how and why planets attained their current orbital positions, why planets differ in size and composition, how and why the atmospheric and surface features of every planet progressively undergo evolutionary changes, and what forces drive planetary evolution. Why does Earth contain the full range of elements from hydrogen to uranium, while only the lighter elements can be found on Jupiter, Saturn, our Sun, extra-solar planets, etc.? How and why did the known extra-solar gaseous giant planets form so close to their central star? Why will water and signs of early forms of life likely be found on each of our nine planets and many moons?

The corroborating evidence is both substantive and substantial. A definitive and testable alternative, the revolutionary FLINE model of planetary origins and evolution is brought full circle by this solution to the proposed Fourth Law of Planetary Motion, now undergoing peer review. The new model reveals the crucial relationship between the spacing of planets and the forces that power their ongoing stages of evolution.

Evolutionary Background of the FLINE Model

The solution to the 400-year mystery proved to be the final link in a revolutionary model of the origin of our Solar System and the evolution of planets — a new concept that evolved during the research years of 1973-1995. This geometric solution that first eluded Kepler in 1595 proved to be the key to bringing the decisive discoveries of the past millennium full circle. It all began with the realization in 1973 that hydrocarbon fuels (gas, petroleum, coal) could not have been created from fossils. The encapsulated imprints of live plants and animals found within coal — the last stage of polymerization of these three fuels — could have been preserved only via a sudden encapsulation of the live plants and animals by petroleum gushing from the earth and inundating lowlands, then polymerizing and cross-linking into solidified coal via these very common processes of Nature. Substantive evidence since then has confirmed beyond doubt that these fuels were created — as was all matter comprising Earth — via these natural processes of Nature, and not from fossils.

The ongoing creation of atomic matter comprising all planets would not be possible without an internal source of energy that keeps the planet active until the energy source is depleted. These internal processes eventually push each planet through five common stages of evolution. Earth is in the fourth (rocky) stage; Mercury and Pluto are in the fifth and final (inactive) stage.

Scientists know that all stars are driven by these internal processes called internal nucleosynthesis (i.e., energy into matter, in accord with Einstein's formula $E = mc^2$). Many scientists also realize that all planets of the Universe are driven via internal nucleosynthesis, often referred to as the heat within. The Four Laws of Planetary Motion (FL) reveal the fiery, dynamic origin of our Solar System. This internal heat, properly called internal nucleosynthesis (IN), is the driving force behind all planetary evolution (E). We have only to look to the skies to observe each of these five stages of ongoing planetary evolution: all at rates in full accord with size.

Fusion-fission processes occur in all spheres of the universe; in making the stars shine, they are the universal powers that drive forward the evolution of all spheres. Rene Descartes, brilliant in his own right, first recorded in 1644 that all planets began as spherical masses of "Sun-like" matter. His writings on the subject appeared 30 years after Johannes Kepler had discovered the last of his Three Laws of Planetary Motion, but had failed in efforts to discover the fourth law explaining the geometric spacing of the six then-known planets. Soon afterwards, Newton discovered gravity holding the solar system together. These great discoveries, along with those of a number of geologists, form the solid foundation of a revolutionary definitive concept of planetary evolution.

The discovery of the mathematical solution to the Fourth Law during the years of 1980-1995 proved to be the final link in the new model of the origin of our solar system and the evolution of planets through five stages of evolution via internal nucleosynthesis. The Four Laws (FL) clearly reveal the dynamic origin of the solar system and how the planets attained their orbital spacing. The processes of internal nucleosynthesis (IN) in full accord with Einstein's formula clearly are the forces that drive all planetary evolution (E) forward until the energy-core source is depleted. These three interlocking and inseparable principles serve as a solid foundation for the revolutionary FLINE model of the planetary origins. This new definitive concept provides crystal-clear explanations that are continually corroborated by every relevant discovery about Earth and all other planets, including the recently discovered exoplanets. The solution to every anomaly of every planets and moon has its taproot deeply embedded in the FLINE model — a statement already proven true by its record of accurate predictions.

The new model remains poised to displace the prevailing, but highly speculative, Accretion concept that has survived via speculative modifications since its introduction in 1796 by Pierre Simon Laplace in his *Exposition of the System of the Universe*. Einstein's work defining the energy-matter relationship that earned him the title of *Man of the Century*, along with Descartes' insight into the nature of planetary cores, Kepler's Three Laws of Planetary Motion and the new Fourth Law, stand as beacons capable of guiding scientists to a fully proven understanding of the origins of solar systems and the evolution of planets — and perhaps to understanding our Universe — in the 21st century.

Sadly, the disbelief in Darwin's animate evolution concept has an equally tragic parallel in the inanimate evolution of all things universal. The two are intimately intertwined; both stem from the same taproot of origins: energy. Nature's greatest universal truth can be summarized in two words: *everything evolves*. Evolution — animate or inanimate — is not possible without an energy source to drive it forward. Every planet is a self-sustaining entity that creates its own compositional features via the internal nucleosynthesis of atomic matter and subsequent polymerization, etc. — including any form of life existing thereon. We do not need to look to outer space for answers; they have always been right before our eyes.

The definitive FLINE model of planetary origins (1973-1999) consists of three chronological and ongoing realities of Nature: (i) the Four Laws of Planetary Motion (FL); (ii) internal nucleosynthesis (IN); and (iii) evolution (E). No solar system is possible without undergoing these inseparable realities that remain interactive until, one by one, they reach their ending. They reveal that planetary evolution is not possible with any type of core other than that hypothesized by Descartes as "Sun-like" — a concept now corroborated by abundant substantiated evidence. Such evidence mounts with each relevant discovery of planetary anomalies, both of Earth and of space probes. From energy mass to inactive sphere, every planet is a self-sustaining entity that creates atomic compositional matter via internal nucleosynthesis throughout its first four stages of evolution.

Whether speaking of the animate or the inanimate, the evidence for evolution is indisputable, incontrovertible, overwhelming and conclusive. To the two revolutions Freud designated as paramount (the ideas of Copernicus and Darwin), we must add another one that is destined to displace prevailing beliefs: the FLINE model of planetary origins and evolution. Had Kepler succeeded in his quest (initiated in 1595) for the Fourth Law of Planetary Motion, there would have been no need for scientists later to establish the speculative, but erroneous, concept of planetary accretion from gases, dust, asteroids, comets, etc. The solution* (1980-1995) to the elusive Fourth Law was the final link in the FLINE model that inevitably, as with the ideas of Copernicus and Darwin, will force scientists to rethink their beliefs about inanimate and animate origins. In view of its factual, non-speculative structure and vast scientific potential, why must this new model face the same hurdles encountered by those revolutionary ideas?

Alexander A. Scarborough

The Five Stages of Planetary Evolution (1)

The Four Laws of Planetary Motion (FL) give vital clues to the dynamic origins of our solar system. But once in their GR orbits, how did the fiery Sun-like masses evolve into the planets we see today?

Descartes, in his *Principles of Philosophy*, defined Earth's interior as being Sun-like (Fig. 2). Later, Buffon, author of the 36-volume *Natural History*, speculated that fragments of the Sun formed into spheres that became planets. Considering the limited knowledge of those times, these two beliefs are amazing in how close they came to an elementary understanding of our planetary origins long before the fiery relationship between nuclear energy and atoms, as shown by Einstein's formula $E = mc^2$, was discovered, and the processes of nucleosynthesis and polymerization were recognized. Einstein's formula and the atomic bomb make it obvious that matter is simply another form of universal energy, and that all atomic matter was, and is, formed from energy. Such transformations of energy into matter, known as nucleosynthesis, must occur, of necessity, under extreme conditions of high temperatures and pressures within masses of energy. Various and varying combinations of these severe conditions determine the types and quantities of atomic elements produced via internal nucleosynthesis (IN) in any specific time frame by stars, planets, moons, and our Sun.

Elements ranging from hydrogen to iron have been detected in our Sun; no elements heavier than iron can be found therein. But heavier elements are produced in other stars and in planets. The more severe conditions created in Earth's thickly encapsulated core results in the production of all the elements, including the heaviest one, uranium.

The evolution (E) of planets (Fig. 3), in full accord with all natural laws, occurs at rates proportional to size: the smaller the planetary mass, the more rapid its transition through each of five common stages of evolution. Astute observations of the planets reveal excellent examples of the five ongoing stages of evolution: e.g., Energy (the Sun), Gaseous (Jupiter, Saturn), Transitional (Uranus, Neptune), Rocky (Earth, Venus), Inactive (or nearly so: Mercury, Mars, Pluto, Moon).

Note that the planets in the picture slowly evolve from hot-to-cold stages in full accord with size and in full compliance with the laws of thermodynamics. The transition from the first stage of evolution into the second, or gaseous, stage begins with the creation of vast amounts of the lightest and easiest-to-make atomic elements, primarily hydrogen and helium, the main elements found in the Sun. Countless numbers of atoms, ions and simple compounds eventually form a thick enveloping spherical blanket high above their source of creation. Simultaneously, as with our Sun, small percentages of atomic matter accumulate on the surface of the hot core. During eons of time, virgin matter billows from the manufacturing core, slowly filling the huge space between the core and the spherical blanket, thereby forming clouds of chemical vapors throughout the towering atmosphere. Scientists see these results in Jupiter and Saturn, and classify them as gaseous planets. Also, all known extrasolar planets, because of their huge sizes, are currently in this second stage of planetary evolution.

In the next phase of transition, the voluminous clouds of atmospheric matter gradually close in on the hot surface, and the precipitation of its chemical vapors is initiated. The precipitates are instantly repelled by the heat; the gigantic battles between the evaporation and condensation forces rage on for eons. Trapped between the heat of the core and the super coldness of space, the virgin matter finally gains a toehold in the form of the first tenuous liquid layers, eventually followed by thin, tenuous crustal formations. Uranus and Neptune should be in this third, or transitional, stage of evolution.

During the fourth stage of planetary evolution, shallow seas cover the entire surface, cooling it sufficiently to permit formation of islands of more permanent, but drifting, crust. The crust thickens as a function of time — an ongoing process for billions of years. Newly created matter from volcanic outpourings of lava, water, gases and compounds of great variety continually build and modify the atmospheres and land systems. The rocky crust grows ever thicker, while the atmosphere grows ever thinner. Earth is a prime example of this fourth, or rocky, stage of planetary evolution.

Planetary energy cannot last forever. When the hot energy core finally is depleted, the planet or moon in question becomes inactive — a dead sphere that cannot be revived. Mercury, Mars, our Moon and Pluto are examples of spheres that are in, or are entering, this fifth and final stage of planetary evolution.

The Five Fundamental Principles of Origins and Evolution of Planetary Systems (Mar 2001).

In 150 A.D. Ptolomy XIII, the brother of Cleopatra, issued an edict placing Earth at the center of the Universe. This belief prevailed for some 14 centuries. Then, in 1543, Nicolas Copernicus published his book, *Concerning the Revolutions of the Celestial Spheres*, which placed the Sun at the center of our Solar System. This heliocentric principle, the first of five highly significant discoveries, initiated the Copernican Revolution that promises a definitive understanding of our Solar System's origin and evolution.

The second principle consisting of the Three Laws of Planetary Motion was discovered in the early 17th century by Johannes Kepler. They revealed (1) that every orbit of the six then-known planets is an ellipse, with the Sun at one focus of the elliptical orbit, (2) that an imaginary line from the center of the Sun to the center of a planet sweeps out the same area in a given time; i.e., planets move faster when they are closer to the Sun, and (3) the squares of their period of one revolution are proportional to the cubes of their mean distances from the Sun. Although Kepler recognized that the six known planets were geometrically spaced, he failed in his attempts to explain this enigmatic Fourth Law of Planetary Motion — a potential Copernican-like revelation of the dynamic origin of our Solar System.

The third fundamental principle, gravity, was discovered by Sir Isaac Newton in the mid-seventeenth century when he saw an apple fall to Earth. He suddenly realized that one and the same force pulls the apple to Earth and keeps the moon in its orbit and, subsequently, the planets in their orbits.

The fourth fundamental principle was brought to light in 1905 by Albert Einstein in the revelations of his famous formula, $E = mc^2$, which was used to work out some of the basic problems of atomic energy. The formula clearly reveals the intimate relationship of energy to matter, which led to the atomic bomb and to the realization that all matter was, and is, continually created from energy.

The fifth fundamental principle was finalized by the author in 1995, exactly 400 years after Kepler's initial attempt to solve the great mystery of why planetary orbits are spaced in an imperfect geometric pattern around our Sun. This solution (1980-1995) to the Fourth Law of Planetary Motion that first eluded Kepler in 1595 conclusively linked all facets of the evolving FLINE model of origins and evolution.

These five fundamental truths reveal a clear understanding of the origin and functions of our Solar System, and the evolution of its planets and moons via internal nucleosynthesis (IN), the transformation of nuclear energy into matter as revealed by Einstein's formula — which applies to all spheres of the Universe during their evolution from energy fireballs to inactive spheres, and accounts for all their nature.

During its quarter-century evolution, the revolutionary FLINE model has compiled an impeccable record of accurate predictions. One example best illustrates this point: Its deadly accurate prediction of the extremely powerful explosions of the 21 fragments of Comet SL9 on Jupiter in 1994 stood alone as a giant among the many puny predictions of others, including those of the world's best supercomputers. Recognition of the true fiery nuclear energy nature of comets made it no contest against the erroneous belief that comets are dirty snowballs (Whipple, 1950), the basis used in the computer models.

The 1975 pamphlet, *Fuels: A New Theory*, the initial publication of the revolutionary concept, led to a number of accurate predictions of fuel reserves — gas, petroleum, coal — that always remained ahead of all other relevant predictions; e.g., the tremendous volumes of methane hydrates were predicted and explained before being discovered during the past two decades. By understanding how these closely related fuels were created via nucleosynthesis and polymerization, one can understand how all planetary systems were, and are, made in ongoing processes that adhere strictly to the laws of physics and chemistry. Their entrapped fossils played no vital part during the creation of these hydrocarbon fuels, nor do they play a vital role in their ongoing creation.

Every planetary sphere is a self-sustaining entity that creates its own compositional matter. The definitive solution to every anomaly of planetary systems has its taproot deeply embedded in these five principles of the FLINE model — a fact that soon should bring the Copernican revolution full circle.

The latest substantiated support of the FLINE model's five-stage evolution of planets came on December 5, 2000, via the news media three days before its scheduled publication in the journal, *Science*. *Researchers find evidence of lakes on ancient Mars* (LDNews via AP) reported that photos from a satellite orbiting Mars suggest the Red Planet was once a water-rich land of lakes (Malin Space Sciences Systems). The photos, taken by the Mars Global Surveyor spacecraft, show massive sedimentary deposits, with thick layers of rock stacked in miles-deep formations. The pictures show clear views of horizontal deposits of rock, a characteristic of sedimentary layers, in the walls of craters and chasms cut into the surface of Mars. Like Earth's Grand Canyon, such formations are possible only in the presence of water. Further, the splitting of planetary crust via either expansion or spotty sinkhole contraction, as well as volcanic actions, must be taken into account as probable causes of other formations.

"We have never before had this type of irrefutable evidence that sedimentary rocks are widespread on Mars," said Michael Malin. "These images tell us that early Mars was very dynamic [as with all active planets — a basic principle of the FLINE model] and may have been [was] a lot more like Earth than many of us had been thinking." Such incontrovertible conclusions, all readily explained by the revolutionary model, continually move scientists ever deeper into the FLINE paradigm, and perhaps, ever closer to its long overdue acceptance as a valid scientific concept.

The anticipated findings on Mars add to the impeccable list of accurate predictions of the past two decades of the evolving FLINE model. This smaller planet is entering, if not already in, its fifth and final stage of evolution (inactive, dead), bleak and barren, just as Earth is destined to become after its current fourth evolutionary stage completes its cycle in a few billion years. Such depletion of the energy core in each planet and moon defines the boundary between the fourth and fifth stages of planetary evolution. These substantiated findings clearly corroborate the fact that planets undergo the five-stage evolution at rates in full accord with size.

The FLINE paradigm goes a giant step further in explaining how the planets evolve through their five stages of evolution, via internal nucleosynthesis from energy into atoms that form the molecular matter comprising all atmospheres, lands, waters, life, etc. Understanding this new model opens the floodgates to full comprehension of these predicted discoveries on Mars, and eventually, to irrefutable solutions to all planetary anomalies. But in the process, scientists must reexamine a number of ancient myths, especially the modified Laplace accretion hypothesis of 1796, allegedly explaining the origin of planets; they do not accrete from gaseous dust, planetesimals and comets, all of which are byproducts of creation processes involving the intimate relationship among black holes, quasars, galaxies, solar systems, stars, etc.

The findings on Mars, Earth, exoplanets and all similar spheres shout out the dynamic origins and ongoing evolution via internal nucleosynthesis of planets and moons. At the current pace of such discoveries of irrefutable evidence on other planets, can it be only a question of time until the shouts are heard and heeded to the point of displacing the antiquated myths about origins of solar systems and the evolution of planetary spheres?

ORIGINS OF SOLAR SYSTEMS AND THE EVOLUTION OF PLANETS

FL-IN-E Model vs. **ACCRETION Model.**

FL-IN-E Model	ACCRETION Model
1. Structured solely with facts. Non-speculative.	1. Remains highly speculative; not factual.
2. Explains the spacing of planets in all solar systems.	2. Cannot explain the spacing of planets in any SS.
3. Explains planetary evolution via internal nucleosynthesis.	3. Cannot explain planetary evolution via metal cores
4. Explains planetary energy sources that drive evolution.	4. Does not explain these energy sources in planets.
5. Explains compositional differences in all planets.	5. Speculates on these compositional differences.
6. Explains differences in sizes of all planets.	6. Does not explain differences in sizes of planets.
7. Explains the sources of moons around planets.	7. Speculates on origins of planetary moons.
8. Explains the origins of planetary rings.	8. Speculates on origins of planetary rings.
9. Capable of explaining all planetary anomalies of our SS.	9. Incapable of explaining our planetary anomalies.
10. Explains origins & differences of all moons.	10. Speculates on origins & differences of all moons.
11. Explains the sources, nature, anomalies of comets.	11. Speculates on these sources, nature, anomalies.
12. Explains sources, nature, anomalies of asteroids.	12. Speculates on these sources, nature, anomalies.
13. Incorporates 3 inseparable principles of Nature.	13. Ignores these 3 inseparable principles of Nature.
14. Explains electromagnetism via natural energy laws.	14. Attributes electromagnetism to metallic cores.
15. Based on works of Descartes, Buffon, Kepler, Olbers.	15. Based on antiquated beliefs of Kant, Laplace *et al*. Ignores works of Descartes, Buffon, Kepler, Olbers.
16. Obeys the Second Law of Thermodynamics.	16. Violates the Second Law of Thermodynamics.
17. Fully explains all anomalies of extrasolar systems.	17. Can't explain any of these anomalies; speculates.
18. Explains why our SS layout is so rare among many SSs.	18. Cannot explain the rarity of our SS's layout.
19. Predicts & explains all anomalies of probes to planets.	19. Cannot predict or accurately explain these results.
20. Has an impeccable record of accurate predictions.	20. Findings are usually surprising & puzzling.
21. Supported by powerful corroborative evidence.	21. Not supported by corroborative evidence.

Alexander A. Scarborough

Accretion or Natural Laws?

The article, *Stars Rise From Ashes in Globular Cluster* (*Science*, 16 March 2001, p 2067) by Govert Schilling states, "Like houses in an industrial area, the stars in a globular cluster are polluted by the exhausts from nearby chemical plants."

"In a paper to appear in *Astronomy and Astrophysics*, Raffaele Gratton of the Astronomical Observatory of Padua and his colleagues describe their discovery of polluted stars in a globular cluster known as NGC 6752, some 13,000 light-years from Earth in the southern constellation, Pavo the Peacock. The cluster is about 100 light-years across and contains millions of stars, which are hundreds of times closer together than the stars in the solar neighborhood." Gratton *et al.* studied the chemical makeup of 18 dwarf stars — stars about the size of our Sun.

Quoting again: "Because the stars in a globular cluster are believed to have formed simultaneously from the same cosmic ingredients, you would expect them to have a similar spectral fingerprint. Instead, the team found huge star-to-star variations in the composition of the stars' outer layers. For giant stars, that wouldn't be too surprising. Their high internal temperatures churn up their insides vigorously enough to carry the 'ashes' of a star's nuclear burning processes from its core to its surface. But that can't be happening in dwarf stars because they're not hot enough. …We'll have to run the [stellar evolution] models to see in detail what's going on."

The main problem here is that the researchers were forced to interpret their findings in the perspective of Poe's Big Bang that, according to many scientists, never happened. These new discoveries are exciting because they offer substantiated evidence favoring the LB/FLINE concept of origins and evolution. Interpretations via the newer concept — even without the necessity of questionable computer models — present a clearer picture of what's happening in dwarf stars like our Sun. The 18 dwarf stars, as with our Sun, do get hot enough "to carry the 'ashes' [atomic elements] of a star's nuclear burning processes from its core to its surface." Every astronomer is familiar with this action of our dwarf Sun in which most elements (almost one-third) up to iron in the periodic table are created and brought to the surface.

Size is not crucial in the actual production of atomic elements in nuclear cores of heavenly spheres; any size star, moon, planet, and comet with an active nuclear core is driven eventually to the inactive final stage of its evolution. Nuclear masses don't need to be massive to create atomic elements; how else could you explain fiery comets, the atomic bomb and nuclear power plants? Of course, the type and quantity of production of elements composing the 'ashes' of a star's nuclear burning processes do depend on size; here a sphere's size is crucial to the production of its specific atomic elements in any given quantities. Slight differences in the sizes of the 18 dwarf stars explain the differences found in the types and quantities of certain elements created in each one. Further, the severe conditions of nuclear cores are attained in other ways; e.g., elements 99 and 100 were formed in the explosion of an atomic bomb.

Powerful ejection from nuclear cores in all active spheres account for the material identified by the researchers as planetary nebulae. But there is no substantiated evidence that our dwarf star, the Sun, ejected planetary nebulae. The Four Laws of Planetary Motion negate the Accretion hypothesis in which planetary rings are said to be leftovers from the planet accretion-formation process rather than ejecta from the ringed planets. The fiery births of the nuclear masses called stars (of all sizes) occur simultaneously with the clouds of gaseous dust in which they usually are embedded; all are ejecta products of larger masses of energy: the Little Bangs concept. Active stars do accrete their surrounding clouds as well as contribute their ejecta to the clouds, a continual exchange until circumstances alter the situation. They cannot initially form from gaseous matter via accretion without violating natural laws; from the time of their birth, they are fiery masses of nuclear energy interacting with those clouds. Pictures of stars in clouds are interpreted more logically in this perspective than via the prevailing Accretion hypothesis. The stars in a globular cluster are indeed polluted by the exhausts from nearby chemical plants — a basic principle of the LB/FLINE paradigm — and stars are the chemical plants. All planets, moons and comets began, and future ones will begin, as small fiery stars that undergo normal evolutionary cycles.

Conclusion

Embracing the great ideas of Copernicus, Galileo, Kepler, Newton, Descartes, Buffon, Dutton, Einstein *et al.*, the revolutionary FLINE model of planetary origins and evolution was brought full circle with the solution to the Fourth Law of Planetary Motion explaining the spacing of planets around our Sun. The original 10 orbits of our Solar System, along with Kepler's First Three Laws, verify both the dynamic origin of solar systems and the correct interpretation of Olbers in 1802 that Asteroids are fragments of disintegrated planets. Further, in view of the acknowledgment by astronomers that each giant gaseous planet in extra solar systems is too close to its central star to have formed there via accretion of dust, gases and planetesimals, adds powerful evidence to the FLINE model that argues strongly against the accretion concept of planetary origins. Thus, this new model seems destined to displace the modified Kant-Laplace Accretion concept of planetary origins early in the 21st century. This change in direction of scientific thought will be expedited with the eventual confirmation of the true nature and sources of comets (ref. example, Comet SL9) and asteroids.

The inseparable connection between internal nucleosynthesis and evolution will lead to understanding the five stages of evolution of planets and moons, and why all active spheres are self-sustaining entities that create their own compositional matter. A prime example of the manner of creation of all planetary matter by means of natural laws is the origin and evolution of hydrocarbon fuels (erroneously called fossil fuels). With its foundation solidified by the Four Laws of Planetary Motion, the FLINE paradigm definitively connects past, present and future discoveries about all planetary systems. The recent discoveries about Mars, entering its fifth (inactive) stage, and other planets and moons continually corroborate, and fulfill predictions of the FLINE model, thereby enhancing its position as a valid scientific theory that warrants closer examination.

At this writing, the new Fourth Law explaining the spacing of planets has been in peer review by The Royal Astronomical Society in London for three years.

REFERENCES

1. A. A. Scarborough, *The Spacing of Planets: The Solution to a 400-Year Mystery.* Ander Publications, LaGrange, Georgia, USA, 1996.
2. W.C. Mitchell, *The Cult of the Big Bang: Was There a Bang?* Cosmic Sense Books, Carson City, Nevada, 1995.

Note: Such original works usually do not require or use many references other than commonly-known theories.

Alexander A. Scarborough

Letters and Memos

To: EOS, AGU, AAAS, Strait, Astronomical Journal, Science News, UGA, Dubuisson, Herndon. Examining the Overlooked Implications of Natural Nuclear Reactors (Herndon, AGU, Eos, 9-22-98)

During the past two decades beliefs about the composition of Earth's core have undergone significant changes. In the early 1980s scientific papers concentrated on how and why a rocky center formed in our planet. Later in the decade the emphasis was placed on an iron core as the more logical composition to generate a magnetic field. The latest concept (Herndon, Eos, Sep 22, 1998) promotes the idea of uranium from outer-space missiles as the core ingredient that furnishes the internal heat via fission alone.

However, all of these theories have problems; all are speculative and contain fatal flaws. For example, a uranium core decaying via fission alone as the source of the heat within would fall far short of the energy necessary to drive planetary evolution, and would provide no valid source for Earth's electromagnetism. We have only to observe our Sun to understand that the lighter elements, primarily hydrogen and helium, created via internal nucleosynthesis (IN) in nuclear masses (the true source of heat and electromagnetism in all active spheres) are the initial building blocks for fusion reactions into all of the atomic elements of which planetary spheres are composed — often including the last and heaviest one, number 92, uranium. Theories based on this specific IN perspective offer more logical solutions to all planetary anomalies.

The most logical concept of Earth's core was first defined by Descartes as being "Sun-like." When one ponders all possible solutions to the mysteries of planetary evolution, this old idea takes on the reality of truth. Some historical information on the subject will fill in the missing pieces of Herndon's article, and perhaps help maintain an accurate record. My initial publications in 1975 on the powerful supportive evidence for nuclear cores in active planets include three titles: *Birth of the Solar System*, *Evolution of Planets*, and *Evolution of Gas, Oil and Coal*. These subjects were first presented in poster sessions at the AAAS Meeting in 1982 in Washington, D.C. and again in the 1983 Meeting in Detroit, Michigan.

The subject was expanded to booklet form in 1977 under the title *Undermining the Energy Crisis*, then to book form in 1979 under the same title — always emphasizing the critical role of nuclear energy cores in the evolution of planets via internal nucleosynthesis of the atomic matter comprising their atmospheres, crust, fuels, etc. During the ensuring years, these revolutionary concepts were presented at various meetings, including the AAAS, the AGU, the AGS, other professional groups and civic organizations. Many abstracts appear in the records of these meetings. However, my manuscripts and news releases continued to be rejected, and on occasions, were handled unfairly by officials in charge of the programs.

So, I was forced to continue writing and publishing in ever-expanding book form as fast as new discoveries were made (and finances permitted it). In all situations, the new knowledge fitted precisely into the ever-evolving concept. These books include: *From Void to Energy to Universe* (1980) (manuscript copyrighted, not published), *New Concepts of Origins: With White Fire Laden* (1986), *The I-T-E-M Connection: How Planet Earth and Its Systems Were Made by Means of Natural Laws* (1991), and *The Spacing of Planets: The Solution to a 400-Year Mystery* (1996). The additional confirming evidence was continually submitted to various organizations and editors, but to no avail except in one case: *Evolution of Gas, Oil and Coal* was published in *Alternative Energy Sources VI* (Vol 3, pp 337-344) by the Clean Energy Research Institute, University of Miami (1983). Unfortunately, a partial version of my *Energy Fuels Theory* was published in the *New York Times* the same day, and thereby became erroneously known as Gold's theory. However, Gold's incomplete concept did not include the nuclear energy source of the elements comprising hydrocarbon fuels — his source was missiles from outer space.

To: Martha M. Hanna, Letters Editor, PHYSICS TODAY, College Park, MD. 06-14-00

Thank you for the belated response to my letter of June 14, 1999 in response to articles by Burbidge *et al.* in the April 1999 issue of your fine magazine.

Just as the Copernican idea of a Sun-centered Solar System revolutionized how mankind views our planetary system, the solution to the new Fourth Law of Planetary Motion (1980-1995) explaining the spacing of planets around our Sun presents a powerful, definitive and irrefutable argument that is destined to change current views in a similar revolutionary manner. This conclusion was reached after more than two decades of researching the truths and myths of science and presenting the evolving factual facets of the FLINE paradigm to various scientific and civic organizations.

During the five years of presentations and submissions since the solution to the new Fourth Law as the final phase of the new concept of origins and evolution, not even one scientist has been able to give a valid scientific reason for its continual rejection. The real reason, of course, is that there is no valid reason other than that its revolutionary nature threatens, and inevitably will displace, current beliefs about how solar systems came into being and evolved into their different stages of evolution.

Over the past twenty years, the FLINE model has an impeccable record of now-verified predictions. Further, it provides a solid foundation for definitively understanding every anomaly of all planets, including the more than three dozen gaseous (second stage of evolution) giants discovered in extra solar systems. As you know, these giants are much too close to their central stars to have formed there via the accretion concept. Simultaneously, this anomaly was predicted and is explained beyond any doubt via the new FLINE model of origins and evolution. The five stages of planetary evolution leave no doubt that evolution is possible only via internal nucleosynthesis that drives it forward: the two are inseparable — one cannot exist without the other.

In comparison, the one-year delay in response to my last letter is not too bad: my manuscript *On the Spacing of Planets: The Proposed Fourth Law of Planetary Motion* has been in peer review at the Royal Astronomical Society, London, for two years and five months. Any comment here would seem superfluous. Meanwhile, I'm continuing to successfully present the FLINE model to scientific and civic groups. The most recent talk was given to a science group at the University of Connecticut on June 5, 2000, and was well received. Further, I'm sending a copy of this letter and my manuscript on the subject to Editor Stephen G. Benka for his evaluation as a publishable paper.

As in the Copernican era, it is time to move onto newer ideas if we are to learn all there is about our planetary systems. When definitive ideas are developed, antiquated ideas, no matter how admired, must inevitably surrender their hallowed places in science. The time for change has arrived.

To: Fred Spilhaus, Jr., Executive Director, AGU, Washington, DC 20077. 10-08-99

Thank you for the letter of October 4 concerning membership renewal. The letter requested me to please let you know why my membership has not been renewed. I will comply, with the sincere hope that some benefit can be derived from a frank response.

During my years of membership and attendance at the meetings, I learned first hand that prevailing beliefs about planetary origins and evolution are well shielded against the intrusion of new ideas and new concepts. Advocates of these beliefs appear determined to retain them at any cost. Press releases advancing their cause are given high priorities, while those in opposition are not permitted into the news media. Old theories are rehashed and retained, and crucial discoveries are permitted interpretations only in the perspectives of these beliefs. Yet nothing fits without more and more speculation.

Discoveries should be viewed in different perspectives in which other interpretations appear to be more logical, more exciting and have more scientific validity. I speak specifically about the FLINE model of planetary origins and evolution that has evolved slowly during the past 26 years into a concept structured solely with basic facts; speculation is not needed. Based on the Four Laws of Planetary Motion (FL) and the internal nucleosynthesis (IN) that drives all planetary evolution (E), this revolutionary concept is aptly named. Without this powerful energy source within every active sphere of the universe, there could be no evolution. This source was first identified by Descartes as "Sun-like", and later by C. Dutton as "that uplifting force", although neither knew about nuclear energy.

Every planetary anomaly has its taproot in that source. All discoveries and theories of Wegener, Hutton, Lyell, Guettard, Cuvier, Brongniart, Hall, Powell, Logan, Dutton and all other geologists can be clearly explained in the FLINE model. All relevant discoveries of the space probes, past, present and future, will fit precisely into the new concept, and can be readily explained thereby. The FLINE model is the only concept that accurately predicted, and now fully explains, the gigantic explosions of Comet SL9 on Jupiter in 1994. Supercomputer predictions were puny in comparison with the actual results, simply because they were based on the erroneous belief that comets are snowballs or iceballs.

The FLINE model is strongly supported by my solution (1980-1995) to the enigmatic Fourth Law of Planetary Motion explaining how the planets attained their spacing around our sun. But until it is permitted to be brought to the attention of the greater scientific community and the public, it will remain dormant until the records are examined by posterity. These records have been preserved throughout the decades of developing this revolutionary model, including a number of abstracts of facets of the concept presented in poster sessions at various scientific meetings with little notice, less publicity and no grants.

At age 76, I grow weary of fighting for openness to new ideas in science, which is the primary reason for dropping my membership in AGU even though I need its support. I would prefer to retain the membership and attend the meetings if there could be some assurance that scientists would be permitted an in-depth examination of these new ideas. Certainly, the FLINE model would inject excitement and new life into the planetary sciences.

To: F. Todd Baker, Physics and Astronomy, UGA, Athens, GA 30605. Aug 28, 2000

The existing model for planetary origins is not consistent with the recent discoveries of nearly 50 extrasolar planets orbiting stars like our own, as described in *A Field Guide to the New Planets* (*Discover*, March 2000). Therefore, I would ask your indulgence for a few moments while I introduce some elements of an alternative model that is consistent with the unorthodox positions of these gaseous giants of which most, if not all, are too close to the central stars to have formed there.

The *Discover* article states that "orbits in which [our planets] were born some 4.6 billion years ago have remained the same ever since. Until recently that was the accepted scenario. But now the detection of extra-solar planets has forced astronomers to re-examine such notions, because they present us with a paradox. Many [if not all] are so monstrous in size, and hug their stars so closely, that they could not have formed in their present positions. The searingly hot stars around which they circle would have melted their rocky cores before they got started. Instead, it's assumed that they coalesced some distance away, then barreled inward over millions of years. And if such chaos characterizes the birth of extrasolar planets, could not similar disorder have reigned closer to home?" However, as previously admitted, these giants could not have formed very far out because of the thinness of the formation substances at greater distances.

This proposed scenario presents more problems than it solves. However, there is an alternative that warrants consideration, one in which the data do fit precisely without the need of assumptions. Such extra-solar planets were predicted and explained by this revolutionary FLINE concept before these amazing discoveries were made. In it there is no need for chaotic disorder in solar systems. The new theory eliminates both the need for gaseous giants to form at much greater distances from their central stars and the need to sling planets out of their orbits in a chaotic manner. It is strongly reinforced by the laws of planetary motion, which include my solution to the enigmatic Fourth Law of Planetary Motion, detailing how the planets attained their orbital positions around our Sun. The solution to this revolutionary law has been submitted to the Royal Astronomical Society in London, where it has been under peer review for more than 2 1/2 years — perhaps a record?

While it is exciting to see these extra-solar systems adding much to the credibility established in past years by the impeccable record of accurate predictions of the FLINE concept, it is discouraging that main stream scientists have been unwilling to attempt, or simply unable, to point out any fatal flaw; perhaps none exists. Researched and developed during the past quarter century, this new theory does not require the speculative assumptions so necessary in other concepts.

I would like to discuss these new findings with you and any associates at your earliest convenience, with emphasis on determining whether or not a flaw can be found in the new concept.

To: The Royal Astronomical Society, Burlington House, Piccadilly, London. Jan.12, 2001
Re: qy016; On the Spacing of Planets: The Proposed Fourth Law of Planetary Motion.

January 12, 2001 marks the third anniversary of submission of my manuscript for review and publication of the proposed Fourth Law of Planetary Motion detailing how the planets attained their spacing around our Sun — the enigmatic solution that first eluded Kepler in 1595. In conjunction with Kepler's three laws, the four laws have proven to be the final link in the revolutionary FLINE model that reveals — beyond any doubt — the dynamic manner in which our solar system and all solar systems came into being and subsequently evolved into their current orbital positions and evolutionary stages.

Researched over a 27-year time frame, this factually-structured, non-speculative model continues to open new avenues of research that inevitably lead to definitive solutions to planetary anomalies that cannot be comprehended fully in the perspective of any other known theory about the origins of solar systems and the evolution of planets. The most recent example, dated Jan 10, 2001: "Two 'clearly bizarre' planetary systems found in the orbits of distant stars are puzzling astronomers and raising new questions about how planets form." (See enclosed ref: *Mystery planets intrigue experts* via the AAS national meeting.) To fully comprehend the two bizarre (and all other) planetary systems, an in-depth understanding of the FLINE model that explains the eccentricities of all planets and answers the 'new questions about how planets form' is crucial. That is why I urge you to add these findings to your long prestigious record of published historical discoveries.

One has only to read *The Spacing of Planets: The Solution to a 400-Year Mystery* to become convinced of its scientific validity and become an advocate of the revolutionary FLINE model of origins and evolution. Herewith are a few testaments from knowledgeable readers. Additionally, information about the book can be found on the Internet at www.iuniverse.com via Bookstore and typing in the author's name.

As with the findings of Copernicus and Kepler, these new concepts are destined to revolutionize the manner in which scientists view the origins and functions of solar systems and planets. Delay of its review and publication denies scientists the opportunities to benefit from its revelations. In view of the tremendous import of the manuscript's contents and in the best interest of scientific progress, I respectfully urge you to expedite the procedure so that scientists may have the opportunities to comprehend their findings in its perspective.

Please advise the status of the manuscript and whether or not there are problems or questions in need of answers that might help in your esteemed evaluations.

To: Alexander A. Scarborough, LaGrange, Georgia 30241. 2001-02-20
From: Amanda McCaig, Editorial Assistant, Astronomy & Geophysics, the Journal of the RAS.
Re: On the spacing of planets: The proposed fourth law of planetary motion - our ref. qy 016

I'm writing to let you know that, regrettably, we are having to find a new referee for your paper. We have done our best to get a report back from the referee with whom we first placed it, but to no avail. I can only keep on apologising, but this, of course, cannot make up for the quite unacceptable delay in dealing with your work. I do hope you are prepared to bear with us a little longer while we find a more reliable, although just as eminent, referee. However, we would quite understand if you preferred to withdraw your paper and place it elsewhere.

To: Dr. Amanda McCaig, Leeds LS18 4NJ, UK   March 1, 2001
RE: Ref. qy016 - ON THE SPACING OF PLANETS: The Proposed Fourth Law of Planetary Motion.

Thank you for your letter of 2001-02-20 advising the status of the subject manuscript submitted in January, 1998 for peer review and publication. Your efforts to get a report back from the referee with whom it was first placed and your efforts now to find a new referee for the paper are deeply appreciated.

You probably are aware of the daunting task you face to accomplish this crucial mission. History teaches that any major breakthrough that poses a threat to prevailing beliefs unfortunately must undergo much resistance from advocates of those beliefs. Perhaps the referee found the paper to be in this category, and was unwilling to commit either a negative or a positive response because of some apprehension of obvious consequences either way.

Revolutionary concepts inevitably run into the major problem of lack of an expert capable of properly judging it, simply because it is a new concept with which no one has become familiar enough to make an accurate assessment because no one is yet an expert in the new arena — a chicken-and-egg situation. Neither has anyone been able to offer a valid rebuttal to any of its intermeshed facets.

I realize its potential impact, and do understand the problem being faced by the manuscript in spite of its factual nature. It was submitted to the world's most prestigious science institution because of your history of properly handling such historical breakthroughs, and I remain prepared to bear with the Royal Astronomical Society as long as it takes to find a more reliable, although just as eminent, referee. A correct assessment of the manuscript demands an unbiased and very courageous person with a strong inclination to understand and accurately judge it; one who realizes that his/her decisive action either way will be recorded throughout the history of science. The concept's impact on the understanding of solar systems should remain as huge and as memorable as the crucial discoveries of Copernicus and Kepler.

Because of its factual nature, mathematical evidence, intermeshed principles, beautiful continuity and continual verification via the ongoing space discoveries, the scientific validity of this 27-year-researched concept seems assured. Because of its tremendous potential for enabling scientific discoveries to be accurately interpreted without the oftentimes fairy-tale aura, I encourage and urge you to push it through the system in the manner that discoveries of this import warrant. The definitive solutions to all planetary anomalies have their taproots deeply embedded in the FLINE paradigm which, in turn, is based solidly on the Four Laws of Planetary Motion and the principles of the inseparable nucleosynthesis-evolution connection; only within its realm can these solutions spring forth crystal clear and beyond any doubt.

Thank you again for your efforts in this highly significant undertaking.

To: Alexander A. Scarborough   April 22, 2001
From: Gordon Rehberg

Received your latest correspondence and all I can say is how completely disappointed I am in the reply from England. It borders on disgusting. You and I will die and disappear from this life and never have the satisfaction of someone realizing the core essence of your work. It's one thing to think people stupid, and another for them to offer a reply such as you've received and confirm the fact.

Was watching a program on *A & E* or *Discovery* the other evening about the planet and oceanographic research. Long session on Methane Hydrates. Bought a 1973 Oceanographic textbook two weeks ago for one dollar at the Dalton Library Book sale. Been plowing thru it. Find many questions in the book published in 1973, have since been answered with research in the past 20 years.

Enclosed are a small series of web pages I have printed out dealing with Methane Hydrates. The web pages go on to infinity with research findings on methane hydrates. Comments like, "Found on the ocean floor, methane seeping from the floor in small discharges along with OIL." Unexplained radioactivity levels at the ocean floor, etc., etc., etc. Unusual life, new undiscovered life, evidently feeding and growing on methane as a food.

I have concluded that most of the people on this planet are not interested in thinking about anything. Most of the scientists cannot question their foundations of knowledge based on theories. Rather, they are secure in thinking these theories are not questionable at all and represent "truths" rather than "theories". It just would drive them out of their comfort zone of basic knowledge, and this they do not question.

I believe the methane hydrate knowledge only further confirms the nuclear core of all planets. It reinforces the concept of the core, the shell, the pressure vessel, the polymerization of compounds, etc. I am not versed in the Chemical Engineering foundations to expand these facts, but know methane hydrates confirm your theories as well as other physical evidence surrounding us. The shear numbers of people that have been exposed to your ideas and the absolute rejection by all of them is confirmation of their primitive mental capacity. I thought two years ago I would have some luck at helping spread your ideas, but must recognize nothing has happened of importance.

Always appreciate your sharing these comments with me. All the best of luck, and hope your health is well and continuing to improve. All well here, and Nicholas is close to reaching "normal", and has decided he would like to teach high school mathematics in Columbia, SC. He will hopefully be able to start this profession in August. Really think he likes the idea of only working 9-10 months a year. Sounds good to me, only 40 years too late.

To: Glenn Strait, Science Editor, The World & I, Washington, DC 20002   04-18-01

Thanks for the interesting and valuable information in the e-mail of March 29, March 30, and April 10, 2001. It's always exciting to get new information on planetary discoveries that seem always to blend so easily into the FLINE model. I would like to comment on the four issues.

The plume activity on Io is as predicted by the new model: it is definitive evidence of the young age of Io, now in the early cycle of the fourth stage of evolution through which all planets evolve. If this moon and Jupiter were the same age, Io would now be an inactive planet much like Mercury. This was discussed in more detail in a previous letter to you.

I'm always amazed at the logic used in such press releases as *Turning Stars Into Gold*. To wit: "Many common elements, such as oxygen and carbon, are known to be made in stars and distributed through the Universe when a star explodes as a supernova. This is the origin of most of the material that makes up Earth [many questions here]. It is becoming clear, however, that normal stars cannot make enough of the heavy elements, such as gold and platinum. Thus, the origin of gold and platinum — on Earth and throughout the Universe — remains a mystery [?]. …When binary neutron stars are close enough to collide, they soon create the most powerful explosions in the Universe." Dr. Rosswog "has found that a large quantity of gold and platinum is made and thrown into space. Relatively small seed nuclei, made of elements like iron, collect neutrons and build themselves up to become heavy elements such as gold and platinum." Rosswog *et al.* "have shown that the relative amounts of elements formed in his computer models of colliding neutron stars match those seen in our Solar System. This provides strong evidence that most of the gold and platinum on Earth was formed in the violent collisions of distant stars." [logic?]

You recall the puny predictions of supercomputers that did not come even close to predicting the super powerful explosions of Comet SL9 fragments on Jupiter in the mid 1990s, thus casting serious doubt on the validity of computer findings and conclusions. The FLINE model alone accurately predicted these powerful explosions, simply because its initial assumptions of comet compositions were accurate. Certainly, heavy metals are formed in powerful explosions, as was illustrated in the explosion of the atomic bomb that created traces of elements 99 and 100. This is indicative of how the full range of elements can be created within the confines of an encapsulated nuclear mass like those found in all active planets that have advanced into their fourth stage of evolution: the reasons reside in the energy core's gradual increases in the internal conditions of extreme temperatures and pressures in which quantities of specific atoms are created at any specific time, eventually to become crustal matter. All planets are self-sustaining entities that create their own compositional matter via internal nucleosynthesis and polymerization. [Ref.: *Fuels: A New Theory* (1975), *Undermining the Energy Crisis* (1977, 1979)].

One basic problem with prevailing theories is illustrated in the news release of April 10: *Giotto's heritage: the past and future of comet exploration*. To wit: "…ESA's Giotto spacecraft made history by obtaining the first close-up pictures of a comet's black, icy nucleus." In reality, close examinations of these photos clearly reveal the true nature of comets: fiery masses of energy encased in a carbonized black shell of its own making, spewing out separate trails of virgin ionized gases and dust through several separate "portholes" of its own making. If comets do not crash or explode first, they eventually become burned-out shells; examples are Phobos and Deimos, the two moons of Mars. Recent findings reveal that comet tails and the gigantic comae remain with comets after they move far beyond the warming effects of the Sun and into the nearly absolute zero temperatures of distant space — exactly as predicated by the new model. Only fireballs have the capability to accomplish these feats.

Glenn, I understand your trepidation about promoting the new model; I felt that way until the data became — and is still becoming — overwhelmingly persuasive. During the 28 years of piecing its volumes of persuasive data together — finalizing it with the solution to the new Fourth Law of Planetary Motion — and presenting the findings to scientific personnel, not a single valid counter-argument has prevailed against it. Nor does one seem likely to do so.

I fully agree with the mainstream model of stellar and supernova nucleosynthesis, but I do not agree with the accretion hypothesis: an illogical concept that cannot be proven, but can be indirectly disproven. I have ventured one giant step further by recognizing the true nature of planetary evolution (E) via internal nucleosynthesis (IN) and polymerization; IN and E are inseparable to the end; one cannot exist without the other. Planetary evolution is absolute proof of internal nucleosynthesis; it is crucial that every active sphere have an internal nuclear engine to drive its evolution. Only in this perspective can all planetary anomalies be understood beyond any doubt; it is the crucial key to understanding origins and evolution. The FLINE model is simple, beautiful, and easy-to-understand: the type of concept preferred by Einstein, who gave us the dynamic formula $E = mc^2$ that reveals the energy-matter relationship, the most fundamental principle of the IN-E concept that eventually led to the solution to the enigmatic Fourth Law.

In 1985-86, a project at Georgia Tech's Nuclear Division confirmed the 3-layer system of the hydrocarbon fuels within Earth's crust, as predicated by the new IN-E model. I discussed IN with Dr. Karam, the Director, who gave no reason to believe that it was not feasible. But I believe his position in the hierarchy prevented his committing to a final answer. The same results came from my visit with Dr. Todd Baker at the University of Georgia's Physics Department, and from the Royal Astronomical Society and the American Astronomical Society with regard to my manuscript on the Fourth Law. This reaction appears to hold true throughout the system with other academic personnel; they prefer not to step outside prevailing beliefs, no matter the data being overwhelmingly persuasive. I understand their position, one much akin to that of the Copernicus-Galileo era. Thus, science has another classic case of an irresistible force of incontrovertible facts meeting an immovable object of highly speculative, unproven, but well-entrenched, beliefs. Who knows the answer to this dilemma? Time, the FLINE's best ally, will tell — but only after the true nature of comets is "discovered" and confirmed. Meanwhile, must students be forced to continue learning beliefs only?

To: Kathy Sawyer, News Reporter, *Washington Post*, Washington, DC 20071   04-30-01

Your article, *Astronomers detect birth of planets* (*Science*, news reprint in the *Atlanta Journal-Constitution*, April 27) was interesting information well done. But some helpful comments seem in order.

The prevailing perspective — in which interpretations of the new scientific data discovered by the researchers are unduly restrictive — is based on beliefs rather than on factual science. The belief that stars and planets accrete from gaseous dust, grains, etc., was initiated by Simon Laplace in 1796. Scientists do not have any definitive evidence of the accretion process, which does have a nice convenience about it. But it raises more questions than it answers: why can't it explain all anomalies of planetary systems?

But there is a definitive alternative that does explain them, one that's based solidly on the Four Laws of Planetary Motion (FL), clearly revealing how the planets attained their orbital spacing around our Sun before beginning their five-stage evolution at rates in full accord with size. Definitive evidence can be ascertained simply by looking to the skies to observe the various stages of evolution of our planets at rates relative to their sizes. From there, it is a simple step into the reality that planetary evolution (E) is not possible without a core engine to drive it forward. It is an irrefutable fact of science: evolution and internal nucleosynthesis (IN) are inseparable; to the end, one cannot exist without the other.

In reality, planets (and all other spheres) must obey the laws of physics and chemistry; beliefs without substantiated facts, and based on misinterpretations of data, are always illusory.

I am writing while under the impression that you are the same young lady who worked for the *Gwinnett News* and interviewed me some ten years or so ago about the erroneous fossil fuels hypothesis. Since then, the fast-mounting evidence finally led to the enigmatic solution to the revolutionary Fourth Law (1980-1995) that brought the LB/FLINE concept full circle. (Herewith are some pages of more details). In spite of the stubborn resistance against it, this new concept is destined to displace current beliefs early in the 21st century; truth based strictly on facts has a way of prevailing in the long run.

One major question to pose to the researchers: How can smoke particles and sand-size grains — typical products of fiery comets — circling a star be interpreted as proof of accretion, especially in view of the powerful evidence that shouts so loudly against their conclusion and in favor of the definitive conclusion that all rings around spheres are composed of ejecta powerfully propelled into orbits around each equator near the end of the second stage of evolution? Saturn, being further along than Jupiter's early second stage, is the prime example of being at the peak of such processes.

The erroneous fossil fuels hypothesis - a costly contributor to beliefs about energy crises - still survives after 170 years as a detriment to scientific progress; unbelievable in this age of enlightenment.

To: Letters Editor, AAAScience, Washington, DC 20005.   June 26, 2001

The article, *Infrared Gleam Stamps Brown Dwarfs as Stars* (*Science* 15 June 2001) concludes that free-floating brown dwarfs form like stars rather than like planets. The conclusion was made based on the belief or assumption that if the dwarfs formed from contracted clouds like a star, a warm, dusty disk should orbit the dwarf and radiate additional infrared light. The additional IR light, if found, would be interpreted as evidence for the presence of a disk.

After finding 63 dwarfs that showed such evidence, Charles Lada *et al.* concluded that since an over-sized free-floating planet formed by agglomeration would not have a disk, these dwarfs must have formed the way stars do. But questions remain: Could these brown dwarfs have the same type of hard-to-detect rings as those recently discovered around Jupiter? Shouldn't the principles of agglomeration remain identical in all sizes of accreted spheres?

The belief that an over-sized free-floating planet formed by agglomeration won't have a disk appears to contradict the consensus belief that all planets and stars form via accretion, usually, but not always, with leftover material from their formation forming into disks. In these cases, how could size make a difference?

Further, "the surprisingly bright IR light from the 63 brown dwarfs in the nearby Trapzium star cluster is helping to make the case that the free-floating brown dwarfs are failed stars and not stray planets." Would the bright IR light from Jupiter and from comets strengthen the case — or would this tend to weaken it? Could these infrared lights be interpreted as coming from a source of hot energy similar to the 30,000°C core of Jupiter (as measured by the Pioneer 10 spacecraft instruments)?

Concerning nuclear energy sources, should size make any difference in the principles involved, whether the smallest source on Earth or inside any active planetary sphere and its moon(s) or comprising the largest stars? Can we be absolutely certain that accretion is the best and only way to create universal spheres? If not, then why not examine other probabilities that meet the criteria of staying strictly within the realm of natural laws that require no speculation? In such cases, size as a function of time readily accounts for all the differences attributed to free-floating brown dwarfs. They were and are made in the same original manner as all other spheres of the universe: via the transformation of energy into matter at rates in full accord with size and in full compliance with all natural laws; i.e., stars, brown dwarfs, planets, moons, etc., all formed, and continually form, and evolve in identical manner.

CHAPTER III

HOW EARTH'S SYSTEMS EVOLVED

"It is nothing short of a miracle that modern methods of instruction have not yet entirely strangled the holy curiosity of inquiry." Albert Einstein

The Myth of Fossil Fuels (1973-1979)

The third and final phase of this FLINE concept describes how planetary systems evolve from energy into the matter that makes Earth such an ideal planet for all of us. The example of the origin of hydrocarbon fuels (gas, petroleum, coal) best illustrates the processes whereby Nature makes all things by means of natural laws. A little background is essential.

During the 1830s, William E. Logan, a graduate of Edinburgh University, managed his family's coal and smelting interests in Wales. Logan's great interest in the origin of coal led him to study some 100 coal beds and seams. In every case he made note of three observations:

1. A bed of bleached clay lay under each coal bed.
2. Within the clay beds were tangled masses of long, slender, fibrous root systems with a thin coating of carbonaceous matter.
3. Well preserved imprints of ferns and other plants were scattered throughout the coal.

Logan concluded that all plants, specifically the Stigmaria ficodes, had turned into coal. Stigmaria structures, some microscopic, some full size, plus larger fossils (branches and tree trunks) were found in the coal. He recorded that "in Stigmaria ficodes we have the plant to which the earth is mainly indebted for those vast stores of fossils fuels."

In the 1920s, J.B.S. Haldane's hypothesis of petroleum created from tiny marine organisms added more credibility to the concept of fuels made from fossils. So it seemed logical that all natural gas is a product of decomposition of plants and animals. Thus, the fossils fuels theory (FFT) became firmly entrenched in scientific literature. Since Logan's time, advocates have attempted, in difficult and costly efforts, to explain their relevant discoveries in its perspective. In reality, as we shall see, much time, frustration and money could be saved if scientists have opportunities to interpret their findings in the new perspective of the FLINE concept.

Contrary-wise, through the years, there have been a number of opponents who kept faith in the belief of non-biological (abiogenic) origins of fuels. The list includes some historical names: Berthelot (1866), Mendele`ev (1877, 1920), Humbolt, Gay-Lussac and others. In more recent times, a significant volume of evidence that argues against the validity of the fossil fuels theory (FFT) has accumulated in the scientific literature. Through intensive library searches, the author has gleaned and condensed much evidence into six critical points:

1. The tremendous volumes and extreme depths of fuels (especially gas) contradict, rather than enhance, the FFT.
2. The patterns of distribution, size and thickness of coal beds and seams simply do not fit into the FFT.
3. No multi-layers of root systems are found in either thick or thin coal beds; carbonized root systems are found only in under-bed clay.
4. Plants yield vegetable oils; coal and petroleum are mineral-oil based.

5. Peat, credited as a transitional stage of some plant-to-coal processes, actually turns into black dirt that retains no imprints of Stigmaria ficodes.
6. The structural integrity of plant imprints can be preserved only through rapid encapsulation of live plants before decay begins.

Each point adds weight to the powerful evidence accumulating against the FFT. For example, today we know that a fossil imprint found in rock was still in its original life form when encapsulated by softer material that later solidified around it. The imprint remained in the rock after dissipation of the encased specimen, which obviously cannot be credited with creation of its encapsulating rock. In this perspective, we can reasonably conclude that the plants whose live imprints were discovered by Logan could not have created the coal; they simply were victims of encapsulation.

Logan erroneously concluded that coal was made from plants. His three astute observations were misinterpreted. In reality, they argue strongly against his hypothesis. Such integrity of plant structure could not have been preserved via the decaying leaf interpretation made by Logan *et al.* Further, his findings can be explained more reasonably by the FLINE concept. Only in this revolutionary perspective can his three observations be interpreted accurately.

Only hot, chemically-contained petroleum could have bleached the clay beds and seeped down the root systems to preserve them via carbonization. Only the sudden encasement of live swamp plants by an encapsulating medium (petroleum) could have preserved their live structural integrity, discovered as imprints of live plants. How did it all happen?

How Hydrocarbon Fuels Formed in Earth's Crust

As confirmed in a research project in Neely Nuclear Research Center at Georgia Tech in 1985-86, the three hydrocarbon fuels (gas, petroleum and coal) that abound throughout Earth's crust, generally exist in overlapping three-layer systems. Coal is the surface fuel, found on the surface or within the top mile of the crust. Petroleum dominates farther down at medium levels, while methane gas is found vastly more abundantly at deeper depths.

At the center of Earth lies its nuclear core, the driving source of the energy that is transformed into the atomic elements that comprise the hot material of the mantle from which the crust is being formed continuously. Carbon and hydrogen are among the most abundant elements created in the nucleosynthesis processes. They are the building blocks of methane gas; the two elements combine to form the vast quantities of methane gas found throughout Earth's crust, much of it in the form of methane hydrates.

During the rocky stage of evolution, virgin elements combine in similar manner to form a large variety of compounds that become the crust of planetary matter comprising Earth's ever-thickening shell. In our fuels example, much of the methane links together, through the process known as polymerization to create other gases: ethane, propane and butane.

When five of the methane molecules linked together, they formed the next product in the carbon-chain series: pentane, the first liquid and the lightest component of light-weight petroleum. Over time, polymerization continued to forge larger and larger molecules of thin and medium weight oils, and finally, thicker crude petroleum. While unknown to Logan *et al.*, scientists now know that all crude petroleum contains tiny particles of coal – the beginning of Nature's final transitional phase from gas to petroleum to coal.

When huge quantities of crude petroleum were forced onto Earth's surface, large areas of swamplands were inundated with hot encapsulating oils that eventually polymerized, cross-linked and solidified as coal.

From this evidence, we can reasonably conclude that the hydrocarbon fuels, gas, petroleum, and coal, should be called 'energy' fuels rather than 'fossil' fuels.

Strong evidence supporting this conclusion abounds. From the scientific literature, the five recognized facts that dramatically reveal the very close relationship of the three energy fuels could be summarized as follows:

1. All wet gases contain lightweight oils (condensate; e.g., pentane), illustrating the first transitional phases via polymerization from gas to petroleum.
2. All petroleum contains gases in varying amounts inversely proportional to the degree of polymerization.
3. All crude petroleum contains tiny particles of coal, illustrating the second transitional phase via polymerization from petroleum to coal.
4. Gummy coals (boghead and cannel) can be classified as either petroleum or coal, illustrating an advanced stage of the second transitional phase (additional polymerization and some cross-linking).
5. Every lump of coal contains oils and gases, illustrating all three phases and the degrees of polymerization and cross-linking of the three fuels, while confirming their close relationship.

A close study of the evidence clearly reveals that gas, petroleum and coal are the first, second and third phases in the formation of hydrocarbon (energy) fuels, and that polymerization is the key to evolutionary changes from gas to petroleum to coal.

The Three-Layer Systems in Earth's Crust

Why would coal, if formed from great swamps of plants and trees, generally be found among and under the rocky layers of mountainous regions like West Virginia? Why are coal beds located at or very near the surface? Why is coal found in veins, many of which are smaller in diameter and relatively short in length? Would not the predominately mountainous distribution pattern of coal and the shapes and sizes of coal veins indicate that great forces pressured gases and liquids into such locations and formations, where polymerization to petroleum and solidification to coal occurred? If so, how could one explain the process, in terms of natural laws, that created these situations?

Why does petroleum exist generally at intermediate depths ranging from near the surface down to 30,000 or more feet? Why is it that the deeper the drilling of wells, the greater the proportion of gas to petroleum? Why should gas be the deepest of the three hydrocarbon fuels? Why do no fossils exist in the deeper oil and gas sites? And how did gas form at such great depths in unimaginable quantities under tremendous pressures at such extremely high temperatures?

Through literature searches in many libraries over the ensuing years, confirmations of the imagined answers to the persistent questions rattling in my brain were sought and found. The excitement of discovering in the literature the facts that substantiate the wild imagination serve to spur one to the next phase of the problem. There can be no turning back or termination of the quest for truth: one can become obsessed with it. This, in spite of the discountenance certain to be encountered along any and all ways that are in opposition to established beliefs. While that can be traumatic, it cannot be a deterrent to progress. History is replete with such examples, and time will never alter the cycle.

To summarize briefly, the energy fuels are arranged generally in three overlapping layers in Earth's crust. Coal is found on or near the surface, while petroleum is found at lower (medium) depths. Deeper drillings result in higher ratios of gas to petroleum. Finally, the deepest drillings may yield only gas. These overlapping layers of gas-to-liquid-to-solid fuels are arranged roughly in ascending order from Earth's interior to its surface by natural laws of physics and chemistry.

The first publication of the three-layered system of fuels in 1975 (*Fuels: A New Theory*) by the author understandably met apathy and some opposition from disciples of the FFT. Additional publications in 1977, 1979, and 1986 managed to gain a few converts, but the numbers remain small at this writing.

Meanwhile, in February 1981, an article on *World Energy Resources*, was published in a special report in *National Geographic*. Included were three maps of the North American continent showing the distribution pattern for each of the three fuels. When superimposed, the three maps revealed very similar distribution patterns for gas, petroleum and coal over the whole continent, thereby adding much credibility to the three-layer concept.

With this background, let's take another look at Logan's three astute observations of the 1830s, and, working in reverse of Nature's processes, continue developing the new theory of formation of these hydrocarbon fuels.

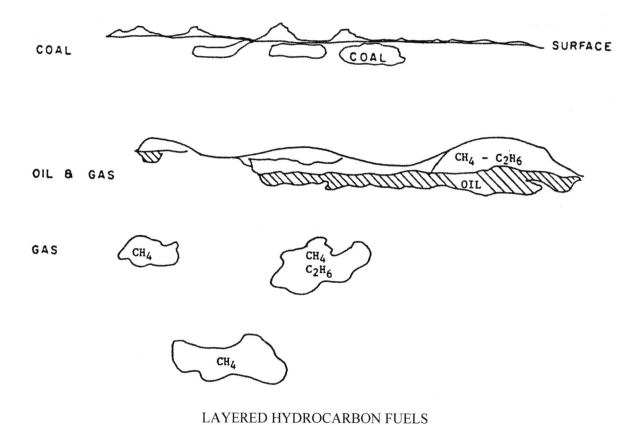

LAYERED HYDROCARBON FUELS

Coal, petroleum and gas are generally layered in the descending order shown.

Figure 6

How Coal Formed From Petroleum

When Logan saw the bleached clay under every bed of coal, he should have questioned why it was bleached. Bleaching is usually accomplished by a combination of certain chemicals at high temperatures. Imagine what might occur if large volumes of petroleum were forced from the ground under thousands of pounds of pressure and at very high temperatures. Such a familiar gusher happened in 1979 in the Gulf of Mexico when an oil well flowed threateningly for nine months before being forcibly closed off.

Hundreds of millions of years ago, such hot oil erupting onto land filled all the low areas (generally swamplands) within reach. It inundated low-lying areas where Stigmaria was usually the dominant plant life. The high heat and chemicals in the petroleum bleached the clay underneath, thereby leaving the first clue for Logan to discover.

By seeping downward and penetrating the root system of plants, the oil preserved the "tangled mass of long, slender fibrous root systems with a thin coating of carbonaceous matter" in their original form. The carbonaceous matter found by Logan was the remains of the oils that had preserved the root systems. If not for this preservative, the root systems would have disappeared rapidly from the scene, as all roots normally do.

Ferns, leaves, branches and even tree trunks were scattered, suspended throughout the lake of hot viscous oil. Thus they were preserved in situ, destined to leave their distinctive imprints, porphyrins and carbonized skeletons in the mass as the oil cooled, thickened, solidified, cross-linked and polymerized into solid coal. Time, temperature and pressure inevitably changed the oil channels and deep ponds into coal veins and coal beds, some as much as hundreds of feet thick. Much of the gas and oil was forced by the tremendous pressures into buried crevices, deep sand formations, porous rocks and strata where it transformed into vast quantities of shale and tar sands.

Thus, the world's coals were created when extremely hot petroleum containing gases from Earth's interior poured out over the swamplands, encasing the plants, seeping down the root systems, and bleaching the under-beds before solidifying into solid coal.

But what could be the source of such vast quantities of petroleum? Could these oils have formed from deeper gases that polymerized?

How Petroleum Formed From Gas

If coal formed from petroleum, then the next step is to find the source of the oils. Nature creates petroleum by the process of molecular chain-building; i.e., methane gas molecules are joined together, much like a chain, by natural processes (polymerization) into longer, larger, molecules known as light oils, or condensate. The lightest of these is identified as pentane, because each molecule contains five atoms of carbon. Each time another molecule of methane is polymerized into the chain, the larger molecule is renamed in ascending order: hexane, heptane, octane, etc. Various proportions of these condensates always are found in all wet gases and in all lightweight petroleum.

As polymerization continues, the oils become heavier and thicker, more viscous. At a higher viscosity, the petroleum becomes known as crude oil. During these transformations, a process called cross-linking is initiated to make ever larger and more rigid molecules. While the term is self-explanatory to many readers, its meaning should be clarified: the molecular chains link together side-to-side, thus becoming less mobile, more solid. Finally, when sufficient polymerization and cross-linking occur, the petroleum begins to solidify into the tiny particles of coal always found in heavy crudes.

Where and why does polymerization of gas into petroleum occur? Starting deep inside Earth, methane gas, under tremendous pressure, seeks the paths of least resistance on its upward journey. Most of it becomes trapped either in porous rock or strata or beneath impervious barriers. Here the high heat and pressure, in conjunction with the contaminants (trace metals) in the gas, initiate and sustain the well-known chemical process of polymerization.

During the past two decades, duplication of the process has been accomplished in a number of chemical labs, and patents have been issued covering the catalytic conversion of methane into gasoline (a mixture of lighter weight oils). The process is not commercially feasible at this time.

A fantastic example of such lightweight oils created by natural processes was discovered on June 5, 1989, near a town south of Riyadh called Hawtah in Saudi Arabia. The gusher tested at 8,000 barrels a day. Later, four more prolific oil wells and one gas well were drilled there. The vast deposits contain rare, super light oil that could be used in a car engine without being refined. The new oil has the color of gasoline and the consistency of water. Its gravity varies from 42 to 49 degrees, as measured by the standard API method. The Hawtah oil contains almost no sulfur or other impurities. The oil field may prove to be the world's largest.

Only a week before Iraq invaded Kuwait, Aramco announced the size of the prospective drilling area: 1,440 square miles. The discovery raises questions that may never be answered: Was this field the next target envisioned by Saddam Hussein? Was it the primary reason he gave up the objectives of the 8-year war with Iran and turned his full attention southward? Was his primary objective the control of oil prices worldwide?

One would expect the Hawtah oils to be free of the contaminants found in crude oils, and they are. Petroleum crudes usually contain numerous substances, both organic and inorganic, including many trace metals, salts, lignite, and coal. Microscopic studies reveal fragments of petrified wood, spores, algae, insect scales, tiny shells and fragments. In contrast to Haldane's theory, it is safe to conclude that these substances are contaminants trapped in the oil; they did not create the oil! These contaminants remain in the mass during and after its transition into coal, thereby often participating in the reactions that create the many chemicals found in coal.

It is important to remember that not all gas polymerizes into petroleum and coal. Molecules of gas are found in every lump of coal and every drop of petroleum. Pockets of these gases in coal mines, when released from entrapment in the black gold, present great dangers of explosion and suffocation.

Source of the Methane Gas

If coal was made from gushing petroleum that had been created from deeper gas, what is the source of the methane? How was it made in such unimaginable quantities under such extreme conditions of temperature and pressure? Is it still being created? The answers to these questions are profound, revealing a logical concept in which the creation of all matter comprising our planet can be explained with very little, if any, speculation.

Before delving into these questions, we need to take a closer look at the vast resources of gas in Earth's crust. Geopressured methane, an example of one type of gas, lies in the untapped gas-laden briny waters buried deep beneath the Gulf Coast. A Baton Rouge hydrogeologist, a leading authority on the subject, calculated the supply at 50,000 trillion cubic feet (tcf) of methane in Louisiana and Texas. The annual use rate of 20 tcf in the USA equates to 2,500 years of reserves in this one location. One tcf is the energy equivalent of approximately 180 million barrels of oil.

In addition to the Gulf Coast reserves, methane is known to exist in a large basin area some 500 miles in length in California, and other states, including Oklahoma, Washington, Oregon and Alaska (250 tcf). These resources can be multiplied many times in the worldwide perspective, since there is no reason to believe that the listed areas have a monopoly on gas supplies. As in past decades, even these estimates will prove far too conservative.

Therefore, are we talking about unimaginable quantities of methane inside Earth – stupendous amounts far too deep and too voluminous to have been made from fossils, or to have come from outer space? What might be a logical source from which such huge volumes could evolve?

One clue lies in finding the source of the carbon and hydrogen atoms that combine to make methane. How and where can these atoms be created?

Einstein and his contemporaries proved that atoms could be forced to release their nuclear energy under specific conditions. This tells us that Nature made atoms from nuclear energy under various specific and extreme conditions, most likely at millions of degrees and fewer than millions of pounds of pressure. Such conditions exist in nuclear masses like our Sun and stars.

Scientists have identified a number of types of atoms on the surface of the Sun, from hydrogen to iron. The glowing hot gases of the Sun are made of about the same chemical elements composing the crust and atmosphere of Earth, including carbon and hydrogen, the building blocks for methane. Since these clues from the Sun suggest the strong possibility that Earth's atoms could have been created from nuclear energy in situ, the key question becomes: Does Earth have a nuclear core that created and still creates the atoms comprising its mantle, crust and atmosphere? This question was answered in the affirmative in Chapter II.

With the concept of a nuclear core in each planet, the enigmatic mystery of planetary origins dropped its impregnable veil, and the dawn of reality came sharply into focus. Too good to be true; something must be wrong. But in perfection, where every piece has fitted precisely into the concept, its first flaw yet remains undiscovered after 27 years of piecing it all together and testing it superficially in scientific meetings.

Evidence for the Making of Abiogenic Methane Gas

Fortunately, it is possible to differentiate between biogenic and abiogenic methane by identifying two isotopes of carbon comprising the gas. Biogenic (biological) methane is enriched in ^{12}C, while abiogenic gas consists of ^{13}C (Hoefs, 1980). The high ratios of ^{13}C to ^{12}C in methane, especially deep gas, indicate abiogenic origins. The high ratios in methane "hot spots" of the Red Sea., Lake Kivu (East Africa), and the East Pacific Rise (McDonald, 1983) suggest abiogenic origins.

Additional background literature by Peyve (1956) and Subbottin (1966) argues that subcrustal abiogenic petroleum migrates up major faults to be trapped in sedimentary basins or dissipated at Earth's surface. Profir'ev (1974) cites the flanking faults of the Suez, Rhine, Baikal, and Barguzin grabens as examples of such petroleum feeders.

This school of thought gained support in later publications; e.g., *The Deep-Earth-Gas Hypothesis* by Gold and Soter (*Scientific American*, June, 1980). The article presented much evidence that earthquakes and volcanoes release gases from Earth's mantle, and such gases may include methane of a non-biological origin. In a later publication, Gold argues that earthquake out-gassing along faults allows methane to escape from the mantle, a process that gives rise to deep gas reservoirs and via polymerization, to petroleum at shallower levels. Further, the *Reader's Digest* (April 1981) published the article *Bonanza, America Strikes Gas*, which tells of geologists hitting field after field of natural gas deep within the nation's bedrock.

These articles and a number of other discoveries and arguments strongly support both the energy fuels theory and its all-embracing concept of internal nucleosynthesis (first introduced in 1975 as the TIFFE concept: The Internal Formation of Fuels and Elements, later given the I – T – E – M concept label). They add powerful support to the one small, lone voice that argued against the hysteria of the energy crisis in the 1970s in attempts to convince the establishment of the true nature of the vast reserves of hydrocarbon fuels in Earth's crust worldwide – a message not heeded until the 1980s.

Today's glutted oil market and reasonable prices attest to the warranted faith in the concept. But even at this writing (1996), the origin of these energy fuels remains erroneously attributed to fossils.

The Rise and Decline of the Abiogenic Fuels Theory

The taproot of the energy crisis of the 1970s can be traced back to the original theories of the 1830s and 1920s: gas, petroleum and coal were made from fossils. This misconception proved to be one of the most tragic scientific myths of our time. Its implications of very limited fuel supplies had traumatic effects during and after the 1970s debacle.

During 1973-1976 the price of petroleum tripled, and fuel shortages resulted in long lines at gasoline pumps, high inflation and soaring interest rates. Some experts predicted that gasoline prices would climb as high as $5.00 per gallon. OPEC seemed in complete control.

It was a time of genuine, irrational fear that fuel reserves would be depleted within a few years. In April 1977, President Jimmy Carter expressed his alarm to a national television audience, declaring the moral equivalent of war on the energy situation. By 1979, oil prices had nearly tripled again, with devastating effects on the economy.

Many causes and effects were debated. Two of the primary reasons were attributed to: (1) the steadily declining exploration and production of petroleum in the USA between 1955-1973, a decline of nearly 70%; and (2) the pervasive fear that the world's fossil fuel reserves were very limited and were being rapidly exhausted. The heightened impact of low supply and high demand, intensified by depletion woes, was traumatizing to many people. Fear and dire predictions ran rampant, at times bordering on hysteria that was duly captured by the news media: experts spreading doom and gloom – all based on the false gospel of very limited reserves of fuels made from fossils.

Much of the sanity that survived remained in relative obscurity. For example, a copy of my 1979 book, *Undermining the Energy Crisis*, was given to Ronald Reagan during his presidential campaign in Louisiana. Its energy fuels concept argued for maximum production of fuels from the vastly underestimated reserves in the USA, while alleviating the unwarranted fear of rapid depletion of the world's supplies. Further, it predicted that the price of gasoline in the 1980s would stabilize at about $1.00 per gallon whenever supplies met demand.

It is not known for certain that he read the book, but immediately after Reagan's election, the new President issued a three-word order to oil companies: Produce, produce, produce. Consequently, the book's predictions proved deadly accurate. What had made it possible to make such precise short and long-term forecasts in opposition to so many prevailing expert opinions? The answer came from understanding how these vast stores of energy fuels were created in Earth's crust — not from fossils, but via the ongoing processes of internal nucleosynthesis and polymerization of these hydrocarbon fuels.

As predicted by the new "energy fuels" theory, and quoting from *Financial World* magazine of November 13, 1990: "...the worldwide glut of oil in the ground, estimated to be over one trillion barrels and rising yearly, will find its way to the market, driving prices inexorably lower." Other estimates run as high as six trillion barrels; even that figure will prove to be much too conservative.

The latest verification of the prediction made by my first publication in 1975 on the origin of unimaginably vast amounts of a biogenic natural gas in Earth's crust appeared in *Science Magazine*, 28 June 1991. In this article, *Fire and Ice Under the Deep-Sea Floor*, Tim Appenzeller describes ubiquitous gas hydrates occurring naturally in deep sea sediments and under the Arctic permafrost in staggering amounts. The total amount of gas hydrates worldwide has been estimated as equivalent to some 10,000 billion metric tons of carbon – twice the carbon of all known reserves of gas, oil and coal.

And who can imagine the vast reserves and the source of the fuel that exists beneath these gas hydrates? Once again, these new estimates of the world's gas reserves will prove woefully inadequate.

The most significant and exciting aspect of these discoveries is the powerful evidence they offer of the manner in which these fuels were made (and are still being made, and will continue to be made throughout future eons, possibly longer than mankind will survive on Earth). These and all future findings will fully vindicate the warranted faith in the 1973 concept of internal nucleosynthesis and polymerization of hydrocarbon fuels.

However, the fossil fuels and condensation/planetesimals/accretion beliefs are firmly entrenched in the scientific literature. They will be very difficult to displace. But time and evidence are definitely on the side of the FLINE model; it does seem only a question of time.

Interest in the origin of natural gas peaked in the mid 1980s, primarily as a result of the annual spring meeting in 1985 of the American Association for the Advancement of Science (AAAS). A session on origins of fuels featured Professor Thomas Gold's excellent presentation of strong evidence for the abiogenic origin of deep methane gas. Attempts made through AAAS channels to permit presentation of my findings since 1973 to reinforce Gold's findings proved futile, in spite of the fact that several of the program's scheduled speakers failed to show.

Although given its due publicity, the concept was destined to fall out of favor because of two reasons. First, the information I could have contributed in the session would have completed the big picture of the abiogenic origins of all hydrocarbon fuels (gas, petroleum, coal). However, Gold's singular attempts to fit his version of the origin in with prevailing beliefs – one being that methane came from outer space and was trapped deep inside during Earth's early formation – were doomed to failure, simply because they are unrealistic, and they simply fail to mesh with facts. Second, through Gold's persuasiveness, a deep well was drilled later in Sweden to test the theory. Since no fuel was found, interest in the abiogenic origin waned considerably.

The concept of the abiogenic origin of methane and subsequent polymerization to petroleum erroneously became known as 'Gold's Theory'.

Until the facts in the situation are sorted out and faith in the concept is restored by some future breakthrough, this vital key to the origins of atmospheric, crustal and internal compounds of all planetary spheres via internal nucleosynthesis now faces an even tougher road to the full credibility it warrants within the community. But this can be accomplished only if scientists are permitted to hear the full, factual story. It is hoped that this book is a giant step along that winding trail to the restored credibility warranted by these findings.

Even without having a chance for a full understanding of the now-completed FLINE model (formerly I – T – E – M) concept, Gold and many others have added significant corroborative evidence to the new energy fuels theory (EFT) since its conception in 1973. Now structured with an ever-increasing multitude of interlocking, incontrovertible facts, the revolutionary concept does indeed seem destined eventually to displace the prevailing erroneous beliefs about the origins of hydrocarbon fuels and other matter comprising our planetary systems.

Revival of the Abiogenic Fuels Theory

Perhaps the breakthrough hoped for above has begun. In the same project study at Georgia Tech (referenced previously), nickel and a number of other metals were found in close association with the hydrocarbon fuels. In the common knowledge that metal catalysts generally play a prominent role in polymerization processes, one or more of these metals seemed to be involved here. The presence of these metals, helium, other gases and oils usually associated with natural gas deposits is explained by the nucleosynthesis processes within Earth. For obvious reasons, such magmatic mixtures are more the rule than the exception.

Additional strong evidence on the abiogenic origin of hydrocarbon fuels was presented by Frank Mango at the August, 1995 meeting of the American Chemical Society. Mango, a research scientist in the geology/geophysics and chemical engineering departments at Houston's Rice University, disclosed that metals such as nickel played a major role in the generation of natural gas.

It all began when the particular light hydrocarbons he was examining "had structures that were fundamentally a contradiction to existing views on the origin of petroleum." Mango's conclusion that "the origin of light [and crude] hydrocarbons could not be the thermal breakdown of biological molecules" is further confirmation of the abiogenic Energy Fuels Theory of 1973.

However, due again to the prevailing belief in the FFT, Mango was thrown off track when he concluded that the nickel promotes "the conversion of decomposing organic debris into natural gas.

Evidence indicates that natural gas forms via the natural affinities of its component elements, carbon and hydrogen, which later polymerize (usually with the aid of metallic catalysts) into even larger and varied molecules of gas and petroleum (light to crude). Then, in reaching the third and final stage, the crude petroleum, always in the presence of numerous other elements, polymerizes, cross-links and solidifies into coal.

The key point in Mango's finding is the further confirmation that <u>the origin of hydrocarbon fuels could not be of a biological nature, but is indeed of an abiological nature</u>. Perhaps this could be the spark that re-ignites the interest of scientists in the true origin of abiogenic fuels: the seed theory of 1973 that eventually grew into the FLINE model during the next two decades.

From Energy to Matter to Life: A Reasonable Continuity.

Scientists have much evidence to show that Nature starts with the simpler elements as building blocks to construct ever-larger molecules. Atoms and molecules with mutual affinities arrange themselves properly and assemble together, under proper conditions, into more complex molecules. Nature's self-assembly process has been recognized by biologists as one commonly found throughout biology. Much like a jigsaw puzzle coming together to make the correct picture, the pieces stick together by hydrophobic and electrostatic interactions, according to David S. Lawrence of State University of New York in Buffalo.

Other laboratories have simulated Earth's primitive atmosphere (methane, nitrogen, water vapor) in experiments yielding all five bases that make up the sophisticated building blocks of the genetic code. Four of these bases, (cytosine, quinine, thiamine, adenine) form DNA, the double-helix molecule that spells out the instructions for all living things. The fifth critical base, uracil, substitutes for thiamine to make RNA, which acts as a master slave to carry out DNA's orders.

The key point here is that chemicals are formed easily by duplicating Nature's primitive codes. Thus, the process of chemical evolution must have been relatively simple. This lends strong credibility to the EFT theme that all matter is built with basic building blocks (atomic elements) under proper conditions of temperature, pressure and time specific to the matter in question.

Located with an underwater sonar system in 1993, a natural gas deposit was discovered 170 miles east of Charleston, South Carolina. Based on preliminary mappings, USGS scientists estimated that the area could contain more than 1,300 trillion cubic feet of methane gas locked up in hydrate deposits. If extracted, that volume would serve the country's needs for more than 70 years, based on 1989 consumption levels. However, extracting gas from the deposit located more than a mile beneath the ocean surface would not be a simple matter.

The deep-sea gas vent is ringed by an unusual formation of mussels, clams and other mollusks. Apparently, these life forms have adapted to their dark gaseous environments by deriving energy from the vented methane and other chemicals. This situation illustrates three important points:

1. The continuity in the chain of dependency in the linked cycles of evolution: the energy-to-matter-to-life relationship in which everything is connected to everything else.
2. The environmental survival principle: adapt or die.
3. These findings are indicative of the huge volumes of abiogenic gas already discovered and the vast deposits yet to be discovered in the crust worldwide.

All three factors are vital aspects of the FLINE concept of an inter-meshing continuity of events in which our SS was created dynamically in the energy form from which our planets and their planetary systems are evolving through five stages of evolution in full accord with the laws of physics and chemistry.

The Universal Law of Creation of Matter

Nature's procedure for making its huge stores of energy fuels can be expressed as follows:

ENERGY to ATOMS to MOLECULES of GAS to OIL to COAL

These processes of Nature are completely reversible by mankind. Beginning with coal, scientists can extract its original petroleum, reduce it to gas molecules, which can be separated into atoms. And as Einstein and the atomic bomb illustrated, atoms can be forced to release the nuclear energy from which they were created.

By expanding this formula to include all matter made from energy, we can derive a simple Universal Law of Creation of Matter (ULCM):

ENERGY to ATOMS to MOLECULES to GAS to LIQUID to SOLID

Planetary cores supply the nuclear energy under various conditions of extreme temperatures and pressures for forging atomic elements, the building blocks of matter. These atoms eventually combine and evolve into countless molecular configurations of Nature's handiworks. To cite a basic example other than methane, the waters of Earth were made by internal nucleosynthesis of the hydrogen and oxygen atoms that subsequently combined to form the countless molecules of water comprising Earth's vast oceans and other water systems.

Internal conditions vary from planet to planet, primarily as a function of core size. The large gaseous spheres composed of lighter elements are products of huge, medium-density, open-to-space cores, while rocky planets containing both lighter and heavier elements are products of smaller, higher-density, encapsulated cores. Additionally, distances of planets from the Sun play a role in shaping their surface characteristics; e.g., frozen gases versus liquid surfaces. Thus, at any given time, planets will differ in composition and outward appearances as functions of size and distance from the Sun.

CONNECTIONS: Gaia, Natural Selection and the LB/FLINE Model.

A review of James Lovelock's autobiography, *Homage to Gaia* (*Science* 9 March 2001) gives some valuable insight into his Gaia concept, along with typical reactions of his peers to the revolutionary theory. To quote the reviewer: "Emerging from Lovelock's interaction with microbiologist, Lynn Margulis, the Gaia hypothesis embraces the notion that Earth's living and nonliving components constitute a set of interactive feedback processes that reflect whole-system scale emergent properties — phenomena not likely to be revealed by disciplinary study of Earth's subsystems alone. Lovelock and Margulis suggested that through these interactions the biota made the physical environment more fit for life, a clear departure from earlier scientific ideas… Gaia is a 'large theory', one further inflated by the claims of some of its advocates that it supplants Darwin's theory of natural selection."

"At the first American Geophysical Union Chapman Conference on Gaia {1988} — itself a controversial meeting because it addressed the topic as a serious science [!! which it is] —geologist and philosopher James Kirchner quoted from various Lovelock and Margulis works over 15 years to demonstrate inconsistent implicit definitions of Gaia." Although Lovelock handled the situation with grace, he has since let his feelings be known, and in *Homage* he correctly calls Kirchner's arguments "sophistry".

The completion in 1996 of the FLINE model initiated in 1973 brought the works of Lovelock and Margulis full circle. This revolutionary concept brings definitive understanding to the connections between Gaia's teachings and Darwin's theory of natural selection. Both have their taproots deeply embedded in the principles of the FLINE paradigm in which the origins and evolution of planetary systems, including its ever-changing environment, are driven by Earth's energy core. Species are spawned continually, and when they eventually cannot adapt to the ongoing environmental changes they become extinct.

While the FLINE model's basic principles are supported by incontrovertible evidence that makes them easy to comprehend, they have yet to win the battle to view Earth as a self-sustaining entity — one that, since its initial orbit around our Sun, has never depended on objects from outer space for the ongoing evolution of itself and its multitudes of systems. Earth's living and non-living components do indeed constitute a set of interactive feedback processes that reflect whole-system scale emergent properties — phenomena clearly revealed in the FLINE model by disciplinary studies of planetary origins and Earth's subsystems.

The connections between Lovelock's Gaia, Darwin's natural selection, and the encompassing LB/FLINE model bring a clearer understanding of John Muir's famous quote: "When we try to pick out anything by itself, we find it is hitched to everything else in the Universe." Muir's great foresight is fully vindicated by the irrefutable connections between these three incontrovertible concepts that bring clear understanding to how and why Earth's systems (and all other planetary systems) function interactively during evolution from energy mass to inactive sphere.

Letters and Memos

To: Gordon Rehberg, Dalton, Ga., October 23, 2000

The article, *The World Has More Oil, Not Less* by Alan Caruba, promotes some truly exciting news. He is doing the world a great service by bringing this critical information out into the open. He and you have accomplished far more in a single year of publicizing the facts about our world's vast petroleum reserves than I have been able to do in the 25 years since publishing *Fuels: A New Theory* (1975). It is very exciting that so many people are beginning to learn the true extent and nature of these hydrocarbon fuels since that first publication explaining the reasons: their abiogenic origins and intimate relationships. Even after a quarter century, most scientists still cling to the antiquated fossil fuels concept. Can you believe that a breakthrough of such tremendous import could be stymied for 25 years — and credited to the wrong person? What has happened to our news media?

Thomas Gold's concept, erroneously predicated on the Big Bang/Accretion concept, was published some five years later than my internal nucleosynthesis-polymerization theory of abiogenic origins of hydrocarbon fuels. Even as late as 1985, when Gold presented his theory in the AAAS Meeting in Los Angeles, he believed, and so wrote, that petroleum came from outer space. At that time, he did not understand the crucial, inseparable connection between internal nucleosynthesis of atoms and their subsequent linkages into all the molecules comprising these fuels and all other matter of Planet Earth. Even though my concept was at least ten years ahead, I was not permitted to be on the same program with him. Gold received big coverage from the news media, and thus the abiogenic concept erroneously became known as Gold's theory. As the record shows, Gold's beliefs in the BB and that petroleum came from outer space proved his undoing when the drilling in Sweden came up dry, causing the abiogenic concept to suffer a severe setback.

Some credibility was regained via the accurate predictions made in my books, *Undermining the Energy Crisis* (1977 & 1979), which foretold vast fuel reserves and why they are continuously being replenished from below via internal nucleosynthesis/polymerization, consequently bringing low energy prices in the $1.00 per gallon range in the 1980s. In spite of its definitive nature and substantial irrefutable evidence, the IN-E concept has yet to be acknowledged by advocates of the fossil fuels hypothesis. In 1995, this revolutionary concept came full circle via the discovery of the elusive solution (1980-1995) to the Fourth Law of Planetary Motion explaining how the planets attained their orbital spacing around the Sun. Together with Kepler's Three Laws, the Four Laws of Planetary Motion (FL) gave closure to the FL-IN-E model initiated by the Energy Fuels concept in 1973-1975. I was happy to hear that after reading my book, *The Spacing of Planets: The Solution to a 400-Year Mystery*, chemical engineers at DuPont proclaimed it the most logical explanation they had ever read. It does seem only a question of time before the new concept inevitably displaces the antiquated fossil fuels hypothesis.

Minus this valuable underpinning, other misconceptions, including the Big Bang, cannot long survive. We are on the verge of a number of exciting breakthroughs to a genuine understanding of origins and evolution. For the sake of our future, you, Caruba and I should continue promoting this irrefutable concept until the full truth becomes more firmly established.

Resources and Ramifications: A perspective of the 1980s.

Most scientists recognize that matter was, and is indeed, transformed from energy particles. But it may be some time before they agree that Earth created its own atmosphere and crust with atoms made in its own internal nuclear furnace. And in forming crust, Earth created (and is still creating) its own energy fuels, beginning with atoms of carbon and hydrogen that combine to create methane, then polymerize into higher gases and petroleum, some of which inundated lowlands, then cross-linked and solidified into coals.

Since these events occurred worldwide, it is reasonable to expect that such fuels can be found almost anywhere in the world if one drills deeply enough in the right places. And since these reserves of energy appear plentiful, and nuclear fusion sources are on the horizon, it is quite possible that mankind will become extinct long before energy fuels are exhausted.

In view of the new concept of creation of hydrocarbon fuels, whole new vistas have been opened to mankind. The probability that these energy fuels might not have the limitations of finite "fossil fuels" has worldwide ramifications. When more widely recognized, the EFT will bring new perspectives to fuel suppliers and to energy-dependent industries, while greatly altering the world's economic and political outlooks for many centuries.

The notion that our supply of energy fuels may be nearly limitless is a certain deathblow to the stranglehold exercised by OPEC in the 1970s and to unreasonably high fuel prices during that unwarranted energy crisis.

In the long-term view, the most popular sources of energy should be natural gas and nuclear fission or fusion. Coal and petroleum, both less plentiful and more slowly renewable than gas, should not be burned, but should be utilized for more sophisticated purposes: the manufacture of chemicals and the assurance of transportation needs.

Certainly, the fear of running out of fuels should never again be a factor in precipitating an energy crisis. The notion of 'fossil fuels" is indeed a fossil of antiquated thinking, the relic of the outdated concept of the 1830s.

Why Earth Will Never Run Out of Oil

To: APG Executive Committee, c/o Editor, AAPG EXPLORER. August 25, 1998

The purpose of this letter is to set the record straight on the concept of abiogenic origin of hydrocarbon fuels and to update you on the latest developments pertaining to this fascinating subject.

The story begins in 1935 with my boyhood realization that a lump of coal is simply solidified petroleum containing pockets of blue-flaming gas, and that dying plants on the farm begin turning into dust at the end of their life cycle — making live encapsulation absolutely essential for live imprints to be found in coal (Logan, 1830s). This inspired a lifetime of curiosity to learn everything possible about how all things came into being. After sufficiently researched evidence finally developed into a viable alternative to the fossil fuels hypothesis and removed all doubt from my mind, I was forced to publish *FUELS: A New Theory* (1975) and *Undermining the Energy Crisis* (1977 & 1979) at my own expense during the energy crisis of the 1970s. In these publications, I accurately predicted the vast stores of hydrocarbon fuels (especially methane) in Earth's crust (and the reasons) and the current low price range of $1.00 per gallon of gasoline when experts were predicting $5.00 per gallon and the depletion of petroleum reserves by 1990.

Unfortunately, some facets of my Energy Fuels Theory (EFT) became known as Gold's theory, which—after Sweden's drilling episode—became unacceptable to the scientific community because Gold's belated efforts failed to present an accurate and total picture of the EFT. Long before the Sweden episode, Gold had refused my offers to work together on the full story to convince the science community of its scientific validity. To compound the problem, my presentations at science meetings thereafter were neither permitted beyond Poster Sessions nor permitted news releases through AAAS and AGU.

Cogent evidence of the abiogenic origin of hydrocarbon fuels and their evolutionary stages from energy to atoms to gas to petroleum to coal, in full compliance with all natural laws, led to the realization that planetary evolution (E) is possible only via an internal source of energy (a.k.a. internal nucleosynthesis) (IN) to drive it forward. In any active sphere, IN and E are inseparable; one cannot exist without the other. From this understanding came another realization—corroborated by ever more powerful evidence—that planetary evolution occurs in five stages common to all planets. To observe the five stages, we have only to look, with insight, to the Sun and planets of our Solar System (SS).

Through the years, these realizations became entrenched in much corroborative evidence. But if planets began as nebulous energy masses, how did they attain their orbital spacing? — a mystery that first eluded Kepler in 1595. Inevitably, curiosity led to the mathematical solution (1980-1995) to the proposed Fourth Law of Planetary Motion detailing how the planets attained their original geometrically-spaced orbits around the Sun and their current, somewhat altered, positions in the SS. Together with Kepler's First Three Laws of Planetary Motion, the Four Laws (FL) bring the FL/IN/E paradigm full circle.

Outer space discoveries continually corroborate the FLINE paradigm; all new evidence fits precisely into the new concept. Examples: the findings on Mars and the data from exoplanets are readily explained in its perspective. If a fundamental flaw does exist in the paradigm, it remains elusive. Yet, since its completion in 1995, it has not been able to get past peer-review advocates of prevailing beliefs, even for Poster Sessions and news releases.

I am confident that the concept's supportive evidence is incontrovertible and that the new paradigm will continue to stand on its merits under the closest scrutiny. In comparisons with prevailing beliefs, the paradigm is clearly superior at explaining planetary origins and evolution—including the abiogenic hydrocarbon fuels—and the role of the inner core in Earth's processes, the generation of the geo-magnetic field and the thermal evolution of Earth (and other planets, moons, etc.). It is a factual, mathematically-proven, definitive and viable alternative to current speculative beliefs about the origins and evolution of planetary systems. Another abstract (herewith) on the subject has been submitted for the AGU Fall Meeting in December under the title of *The FLINE Paradigm: Critical Insights into the Origins of Solar Systems and the Orbital Spacing and Subsequent Evolution of Planets*. After three years of rejections, I suspect that my skepticism of the objectivity of reviewers will again be confirmed.

But since this definitive paradigm is crucial to understanding the origin and evolution of hydrocarbon fuels (and all planetary matter) via Earth's internal processes, can you suggest how we might work together, in the best interest of science, to have these findings brought into open discussions among AAPG members? Its potential benefit to science appears unlimited.

To: Glenn Strait, Science Editor, *The World & I*, Washington, DC. 3-21-99

Being familiar with my work of the past 26 years, a local geologist, Grady Traylor, sent me a copy of the exciting article, *It's No Crude Joke: This Oil Field Grows Even as It's Tapped* (*Wall Street Journal*, WSJ, April 16, 1999) by Staff Reporter, Christopher Cooper. The results described therein add more absolute proof of the *Energy Fuels Theory* (EFT) that has been the backbone of my copyrighted writings since 1975, the year of the initial publication entitled *Fuels: A New Theory*. First named the TIFFE (*Theory of Internal Formation of Fuels and Elements*) in my book, *Undermining the Energy Crisis* (1977, 1979), it was re-titled *The Energy Fuels Theory* in 1980, and led inevitably to the concept of the five stages of evolution of planets, which, in turn, led to my pending *Fourth Law of Planetary Motion* (1980-1995) explaining how the planets attained their orbital spacing around the Sun.

Together, the Four Laws of Planetary Motion (FL), internal nucleosynthesis (IN) and evolution (E) form the FLINE paradigm (1973-1996) of the origins of solar systems and the evolution of planets. Thus this new concept, 23 years in the making, consists of three chronological, inseparable and ongoing realities of Nature. The WSJ article is the first recognition by scientists of the initial phase of this revolutionary concept that openly predicted and fully explains its findings. The rest of the story is even more exciting.

Cooper's article is timely in that it comes on the heels of the discovery of a multiple-planets solar system in the Upsilon Andromedae system (AAS news release, April 15, 1999). The findings in both articles add powerful corroborative evidence to the FLINE paradigm—which predicted and explains these discoveries. Although many facets of this revolutionary concept have been presented in a number of national science meetings (including the AAAS in Anaheim on January 23, 1999) and submitted in news releases, the FLINE paradigm remains relatively unknown within the scientific community. Now beyond the shadow of doubt, it is *the crucial key to understanding the origins and evolution of planets*.

As with many historical recordings, the facts in Cooper's article need some revision to set the record straight. Some five years after my initial publication (*Fuels: A New Theory*), I read *The Deep-Earth Gas Hypothesis* by Thomas Gold (*Scientific American*, 1980). I wrote and offered to work with him on the subject; he refused. In the 1985 AAAS Meeting in Los Angeles, Gold and I clashed on different versions of the abiological origins of these hydrocarbon fuels. The record will show that Gold believed at the time that Earth's petroleum was brought from outer space in meteorites, as opposed to my nucleosynthesis version (1975) of its internal creation. He did not delve into the origin of coal, the final product of evolution of hydrocarbon fuels. Because he did not understand the full IN concept, Gold failed in future efforts to find oil in Sweden; consequently, the concept lost credibility—until now. Cooper's article restores this credibility—and that is why it's historical and exciting. I believed that I would not live long enough to see this happen. In spite of my ceaseless efforts and Gold's publicized failures, the WSJ and others understandably and erroneously credit Gold with the concept of abiogenic fuels.

In view of these corroborating discoveries, the FLINE paradigm becomes even more irrefutable. Now more than ever, it appears destined to displace prevailing dogma about origins of solar systems and evolution of planets and fuels. Eventually, it should rank high among the most monumentally significant discoveries of all time. And it guarantees us a vast supply of hydrocarbon fuels for a very long time.

To: Dr. John Chappell, Jr., San Luis Obispo, CA 93406. April 22, 1999

The enclosed article by Christopher Cooper appeared in the WSJ two days after presentation of my paper on the FLINE paradigm at the NPA-SWARM Meeting in Santa Fe. If published a few days earlier, it would have been used as additional corroborative evidence vindicating this revolutionary concept that began with the EFT in 1973 and came full circle with the pending Fourth Law of Planetary Motion in 1995. For sure, it strengthens my faith in every aspect of the FLINE paradigm. I consider its supporting evidence incontrovertible. But this was always the case, since my research during the past quarter century concentrated on the separation of the facts and myths of science.

The paradigm both predicted and now explains the anomalies of Eugene Island 330's "oil field that grows even as it's tapped." The sub-headline states: "odd reservoir off Louisiana prods petroleum experts to seek a deeper meaning." To get the final answers spelled out in the paradigm, they will indeed have to go all the way into the energy core to understand the origins and evolution of planets and all planetary matter, including the hydrocarbon fuels. Their eventual understanding of internal nucleosynthesis finally will dispel the erroneous belief in the Big Bang concept in which all elements were created in the very beginning — an added bonus that will please Bill Mitchell.

In all probability, the next breakthrough can come only if scientists are permitted to learn about the FLINE paradigm. My only hope to reach them is through the next AAAS Meeting in 2000 and a subsequent news release that is permitted to be picked up by the news media. Otherwise, I know of no editor willing to stick out his/her neck to risk that challenge to prevailing beliefs. Do you think that Stephen Gould could be influenced to permit the proposed symposium to be on the AAAS program in Washington in 2000?

I'm enclosing some additional information that is self-explanatory.

To: Dr. Floyd E. Bloom, Editor-in-Chief, *Science* Magazine. April 28, 1999

In response to your challenging editorial *Think Ahead* (*Science,* 2 April 1999), I submit herewith the information concerning a revolutionary concept that predictably will guide scientists to total knowledge of the origin, orbital spacing and evolution of planets. I predict that by 2050, every planetary anomaly will have been solved, conclusively proven and taught as standard textbook material — all within the realm of the FLINE paradigm.

This revolutionary concept began in 1973 with substantive evidence against the fossil fuels theory — the same evidence that pointed precisely to a more logical theory strongly supported by natural laws of physics and chemistry. First published in 1975 under the title of *Fuels: A New Theory*, it was presented in Poster Sessions at the AAAS Meeting in Washington, DC in 1982 and again at the AAAS Meeting in Detroit in 1983. Its key feature was, and remains, the evolution of hydrocarbon fuels via internal nucleosynthesis and polymerization: from energy to atoms to molecules of methane that, via polymerization, evolve into higher gases, light oil, crude petroleum and finally into coal. The rapidly mounting corroborative evidence was presented at various science meetings during the next decade, but to no avail. In 1985, Thomas Gold was permitted a full-day symposium on the subject in the AAAS Meeting in Los Angeles. My offer to work with him to get the full message across in a convincing manner was refused. As you know, the consequences were devastating; the concept lost credibility.

The *Wall Street Journal* article *It's No Crude Joke: This Oil Field Grows Even as It's Tapped* (WJS, 16 April 1999) is the most recent corroborating evidence that should serve as the final proof of the *Energy Fuels Theory* of 1973 that both predicted and explained this anomaly.

The mounting evidence inevitably led to the realization that all planets are self-sustaining entities that create their own compositional matter via internal nucleosynthesis, thus accounting for their five common stages of evolution. Planetary evolution would not be possible without an internal source of energy to drive it forward. Simply put: no energy source, no evolution. The corroborating evidence continues to mount. Every anomaly of planetary systems is traceable back to its source of internal energy.

This awareness inevitably led to the proposed Fourth Law of Planetary Motion (1980-1995) explaining how the planets attained their orbital spacing around the Sun. Now complete, the FLINE paradigm is comprised basically of three chronological, inseparable and ongoing realities: the Four Laws of Planetary Motion (FL), internal nucleosynthesis (IN), and evolution (E). Every anomaly of solar systems can be understood within the realm of this revolutionary concept. Yet, for inexplicable reasons and to the detriment of science, it remains relatively unknown within the greater scientific community.

But by 2050 this scenario of the unknown will be, of necessity, a thing of the past.

To: Thomas McMullen, Georgia Southern University, Statesboro, GA. 5-15-99

Thanks for the interest expressed in your letter of May 12 concerning the background of my *Energy Fuels Theory* (EFT) of 1973. I'm grateful for this opportunity to set the record straight. It is a long story, but I will be as brief as feasible.

The initial sparks can be traced to 1935 when my seventh-grade teacher introduced us to the fossil fuels concept that gas, oil and coal came from plants and animals. This contradicted my observations that plants and animals on the farm quickly turned to dust after dying. And when we burned coal in fireplaces in those tough times, the small blue jets of flames were recognized as emanating from trapped pockets of gas. The rainbow colors found inside lumps of coal seemed symptomatic of oil. And the tarry matter therein was chewed to whiten the teeth. That was the year I decided to concentrate on science, specifically chemistry, to learn the truth about these fuels and eventually, of how everything came to be as it is. My curiosity would not be denied.

Immediately after graduation from UGA (BS in Chemistry), I was drafted into the army, and subsequently married, earned another degree (Chemical Engineering at Georgia Tech) and went through the usual struggles to earn a living while raising three children. In March of 1973, the emerging energy crisis brought my thoughts sharply back into focus on the true nature of the hydrocarbon fuels erroneously known as fossil fuels. All the pertinent knowledge and answers that had been accumulating in my sub-conscious since 1935 suddenly came together in a flash: these fuels were intimately related; coal was solidified petroleum, which previously had polymerized from basic methane gas formed deep within Earth.

These thoughts, backed by more research findings, were first published in pamphlet form in 1975 under the title *Fuels: A New Theory*, followed by an updated version later in the year. In 1977, the concept was expanded to the 40-page booklet, *Undermining the Energy Crisis*.

The next expansion was a 140-page hardback book in 1979 under the same title *Undermining the Energy Crisis*. Since I was unable to obtain any financial aid, or interest any editors, all were copyrighted, published, advertised and sold at personal expense throughout the USA, primarily to libraries. A 1979 copy was given to Ronald Reagan's presidential campaign manager during their swing through Louisiana that same year. After Reagan's election in 1980, his first edict to the oil companies was "Produce, produce, produce." They did, and the energy crisis soon abated. The experts' predictions of running out of fuels by 1990, based on finite resources of the fossil fuels hypothesis, proved erroneous (as predicted), while all other predictions of the EFT have proven true — yet the EFT has gained little scientific credibility.

In June, 1980, *Scientific American* published an article, *The Deep-Earth Gas Hypothesis*, by Thomas Gold and S. Soter. While I suspected that they had gotten the idea from one of my books sold in the Ithaca, NY area, I had no proof of it. I wrote Gold about my publications in this field, and offered to work together to get the full concept of the EFT into the scientific literature. He did not respond. My first presentation of the completed fuels concept to an international audience of scientists occurred in 1982 in a Poster Session at the AAAS Meeting in Washington, D.C. Backed by my employer, Ralph Howard, President, Kleen-Tex International here in LaGrange, we distributed pamphlets entitled *New Concepts of Origins*, which contained a brief synopsis of each of its three compositional subjects: *Birth of the Solar System*, *Evolution of Planets*, and *Evolution of Gas, Oil, and Coal* (the EFT). This was repeated the next year at the AAAS Meeting in Detroit. In 1985-1986, Ralph sponsored a research project at Georgia Tech confirming the EFT's three-layer system of hydrocarbon fuels in Earth's crust. As the corroborating evidence continually mounted, facets of the concept were presented to various organizations of scientists, military and civic groups during the next 13 years. You mentioned the last meeting in Anaheim in your letter. As the Chinese would say: Very long time, very little headway.

In 1985, Gold had a full-day symposium on the subject of origins of fuels at the AAAS Meeting in L.A. Again, I offered to work with him: again he refused. While he made the headlines, the concept was set back more than 14 years when the oil well Gold convinced Sweden to drill through granite came up dry. But the abiogenic concept became known erroneously as Gold's theory. In reality, the only point on which he and I agreed was the polymerization of gas into petroleum. For a reason I did not comprehend, Gold believed that petroleum came to Earth on meteorites, one of which cracked the granite drilling site in Sweden. This was the reason for drilling in that specific spot. To my knowledge, he never did delve into an explanation for coal or into the deeper subject of internal creation of the gases. He did not appear willing to grasp the full significance of the EFT, and remained aloof from it while clinging to facets of the prevailing beliefs —perhaps to minimize possible adverse reactions from other scientists, or possibly to avoid direct conflict with my earlier copyrighted material.

My hydrocarbon fuels concept in the 1970s was first called the TIFFE (Theory of Internal Formation of Fuels and Elements). My 1986 and 1991 publications called it the ITEM (Internal Transformation of Energy to Matter) theory. In the 1996 book, *The Spacing of Planets*, it was changed to FLPM/IN (Four Laws of Planetary Motion/Internal Nucleosynthesis). This change came about to include the manner in which the nebulous planetary masses of Sun-like energy were spaced in orbits before beginning their five-stage evolution into the planets we see today. This was made possible by my geometric solution to the proposed Fourth Law of Planetary Motion (1980-1995), which brought the original 1973 concept full circle. Since then, it has evolved into the FLINE model of planetary origins that includes the five-stage evolution (E) of planets, which is the inevitable result of internal nucleosynthesis (IN). The two (IN & E) are inseparable and ongoing, along with the Four Laws of Planetary Motion (FL). Although the FLINE model was detailed fully in the 1996 book, this full acronym only came along soon after that publication.

Meanwhile, efforts to obtain a grant and to get the concept, or facets thereof, into scientific publications during the past 24 years always proved futile. But many of my abstracts did make it into published science programs, which gave some assurance that these crucial discoveries would be preserved for posterity. My primary concern was to not let them die on the vine before someone in an influential position would be willing to examine their substantial corroborating evidence. The main problem, as I saw it, was that my work was being rejected by peer-review advocates of prevailing beliefs that would be seriously challenged by these new ideas. Who could blame them? But through the years, the reviewers were never able to offer a valid rebuttal or a valid reason for the rejections.

The solution to the proposed Fourth Law of Planetary Motion was submitted to The Royal Astronomical Society in London in mid-January of 1998. Their only response has been the acknowledgement of its receipt and the assignment of the ID number qy016. The AAS Journal quickly rejected it in 1998, again without a valid reason. However, the manuscript, *The FLINE Paradigm: Definitive Insights Into the Origins of Solar Systems and the Orbital Spacing and Evolution of Planets*, requested by the science editor of a magazine in Washington, D.C., was recently mailed to him. Hope springs eternal. If published, I believe it will have an impact at least equal to that of the ideas of Copernicus and Kepler.

Thanks again for your valued interest. If I can be of further assistance, please let me know.

Re: News Release (Via the AAS; Rejected). For Release: May 19, 1999

On consecutive days in April, two exciting and highly significant news items were published. The initial item, a news release via the AAS, revealed the discovery of the first multiple-planets solar system other than our own. The second news item was headlined *It's No Crude Joke: This Oil Field Grows Even as It's Tapped* (*Wall Street Journal*, April 16, 1999). Strange as it seems, the two discoveries are intimately linked and that linkage, coincidentally, had been presented in a science meeting in Santa Fe on April 14 by Alexander Scarborough in a paper titled *The FLINE Paradigm: Definitive Insights Into the Origins of Solar Systems and the Orbital Spacing and Evolution of Planets*. Scarborough, age 76, is a retired research chemist now with Ander Consultants in LaGrange, Georgia.

The news release about the distant solar system tells of three giant planets orbiting a star. The innermost planet contains at least three-quarters of the mass of Jupiter and orbits only 0.06 AU from the star. The middle planet contains twice the mass of Jupiter and orbits at 0.83 AU from the star. The outermost planet has a mass of at least four Jupiters and orbits at a distance of 2.5 AU. "The gaseous nature of the planets and their orbital positions provide powerful evidence against current beliefs about how solar systems form. These findings will encourage scientists to alter their beliefs about such formations occurring from dust, gas, planetesimals, comets, etc.," according to Scarborough.

Conversely, the findings fit precisely into the FLINE model of planetary origins featuring Scarborough's mathematical solution to the proposed Fourth Law of Planetary Motion, explaining the spacing of planets in solar systems. The masses, velocities, distances and gaseous nature of the three huge Jupiter-like planets account for their orbital positions and their being simultaneously in the second stage (gaseous) of the five common stages of planetary evolution. "The FLINE model consists of three chronological, inseparable and ongoing realities: the Four Laws of Planetary Motion (FL), internal nucleosynthesis (IN) and evolution (E). In every solar system, each reality is dependent on the other two," Scarborough stated. Although single- and multiple-planets systems apparently are abundant throughout the Universe, solar systems with nine planets that include an Earth-like planet and three distinct Asteroids belts like ours should be very rare. This is assured in the FLINE model by the strict prerequisites imposed by mathematics and natural laws on the manner in which these systems are created.

The exciting article about the oil field that grows even as it's tapped describes how very hot oil is being forced under very high pressure up through the bottom of the oil field it is refilling — precisely as predicted and explained by Scarborough's *Energy Fuels Theory* (EFT) (1973-1979). The hot petroleum, a polymerization product from virgin methane made deep within Earth, continuously seeks the upward paths of least resistance in its journey towards the surface. "This crucial facet of the FLINE model, supported by a large amount of incontrovertible evidence garnered during the past 26 years, exposes the obsoleteness of the antiquated fossil fuels hypothesis that originated in the 1830s, and this new finding is a powerful addition to the many corroborating facts comprising the EFT," Scarborough added. "The abiogenic origin and evolution of fuels is an example of the typical manner in which all planetary matter was, and is, created via internal nucleosynthesis within every planet."

These findings, along with the discoveries about Mars' tectonic plates, magnetic stripes and its surface features, were predicted, and are explained readily, by the FLINE model. This paradigm clears the path to understanding planetary origins and evolution, and it guarantees us a vast renewable supply of fuels (gas, petroleum, coal) for a very long time. More scientists are beginning to interpret their findings in its perspective. "It is a factual, non-speculative, easy-to-understand concept that makes a lot of sense," according to Dr. Morton Reed, a former professor of thermodynamics at Auburn University. "It provides answers to scientists who rethink current beliefs about planetary origins and evolution."

To: Dr. William Mitchell, Institute for Advanced Cosmological Studies, Carson City, Nevada 89702.   June 12, 1999

Thanks for the tip on the article, *Why We'll Never Run Out of Oil* (*Discover*, June, 1999) by Curtis Rist. I finally was able to get it downloaded from the local library's computer. Rist did an excellent job in describing the vast quantities of petroleum reserves that have continually increased ever since the headlines in 1935 first warned of running out of oil by 1960, and his information closely agrees with my early predictions and publications (1970s) on the subject. His reasoning of improved technology for continually discovering new sources and salvaging oil from abandoned fields does ring true; however, it is only a small part of the big picture.

In the latter part of the article, Rist makes the mistake of going along with the antiquated fossil fuels hypothesis (FFH) to explain the source of gas, petroleum and coal. Perhaps that is the reason he did not mention the discovery in the early 1990s of the world's largest oil field located in central Arabia. The field contains only lightweight oil slightly heavier than gasoline, and is close to being suitable for use in automobiles. This gigantic field of almost-refined oil could not possibly have been created via the FFH. At the same time, it amounts to incontrovertible evidence against prevailing beliefs about creation — not only via the FFH, but also via the Accretion aspect of the Big Bang.

Simultaneously, the lightweight oil field in central Arabia is incontrovertible evidence for the FLINE model of creation in which all matter evolves via internal nucleosynthesis (IN) from hot to cold and all planets evolve through five common stages of evolution (E). The latter is strongly backed by the Four Laws of Planetary Motion (FL); all are backed by a vast amount of substantive evidence accumulated and self-published during the past 26 years. More recent confirmations of the FLINE model and its accurate predictions include the findings on Mars and the discovery of the oil field that grows even as it's tapped. If any valid argument could be made against this FLINE model, shouldn't someone have been able to find at least one during the many years of its evolving facets?

While Rist is correct in figuring that we will never run out of oil, where was he in 1973-1980 when I needed support for the fledgling Energy Fuels Theory (EFT) that accurately predicted and definitively explains the source of these apparently inexhaustible hydrocarbon fuels? For scientists to continue promoting current dogma on planetary and universal origins in apathetic opposition to the overwhelming evidence favoring more factual models does a great disservice to science, to students and to the public.

Just today I received another annual rejection from Dr. Carlyle Storm, Director, Gordon Research Conferences, concerning my application to attend the 1999 Conference on *Origins of Solar Systems*. Can you believe such a Research Conference could have a valid reason to refuse to critically examine the solution to the greatest mystery of our Solar System: the long-sought solution to the proposed Fourth Law of Planetary Motion explaining the orbital spacing of planets around the Sun? Copernicus was fortunate not to be living in these times. It is tragic that scientists are denied the right to debate issues crucial to the advancement of science, even sadder that they are denied opportunities to hear the rest of the fascinating story whereby they could interpret their findings in more sensible perspectives. But as Max Planck concluded, science advances one funeral at a time.

From: Glenn Strait, Sent: Wednesday, July 07, 1999 8:43 AM,
Subject: FYI update.437 (fwd)

This entry in a collection of short news items put out by the American Institute of Physics describes the discovery of and theorizes about minute traces of radioactive iron in the bottom of the ocean. Would IN theory claim IN of this iron? If so, how would you explain the very limited concentrations of the iron?

PHYSICS NEWS UPDATE

The American Institute of Physics Bulletin of Physics News Number 437 July 2, 1999 by Phillip F. Schewe and Ben Stein.

Star Material Discovered In South Pacific.

Interstellar matter formed in a supernova has been discovered on Earth now for the first time. Light coming to Earth from distant supernovas is recorded all the time. Likewise, a dozen or so neutrinos from nearby Supernova 1987A have been detected. But atoms from supernovas are a different matter. In a sense, all the heavy atoms on Earth have been processed through or created in supernovas long ago and far away. But now comes evidence of atoms from a supernova that may have been deposited here only a few million years ago. An interdisciplinary team of German scientists from the Technical University of Munich (Gunther Korschinek, 011-49-89-289-14257, the Max-Planck Institute (Garching), and the University of Kiel have identified radioactive iron-60 atoms in an ocean sediment layer from a seafloor site in the South Pacific. First, several sediment layers were dated, and only then were samples scrutinized with accelerator mass spectroscopy, needed to spot the faintly-present iron. The half-life of Fe-60 (only 1.5 million years), the levels detected in the sample, and the lack of terrestrial sources point to a relatively nearby and recent supernova as the origin. How recent? Several million years. How close? An estimated 90-180 light years. If the supernova had been any closer than this, it might have had an impact on Earth's climate. The researchers believe traces of the Fe-60 layer (like the iridium layer that signaled the coming of a dinosaur-killing meteor 65 million years ago) should be found worldwide, but have not yet been able to search for it. (K. Knie *et al.*, Physical Review Letters, 5 July 1999.)

To: Glenn Strait, Science Editor, The World & I, Washington, D C. July 14, 1999.
Subject: American Institute of Physics, Bulletin of Physics News, #437 July 2, 1999.

Thanks for the e-mail news entitled *Star Material Discovered in South Pacific* which describes the discovery of and theorizes about minute traces of radioactive iron in the bottom of the ocean. First, let me briefly answer your two questions: Yes, the IN phase of the FLINE model does explain this iron and its very limited concentrations. One of the basic claims of this revolutionary concept is that all planets are self-sustaining entities that create their own compositional matter via internal nucleosynthesis (IN); they do not depend on speculative claims that matter came from outer space, or that planets accreted from gas, dust, comets, asteroids, etc. The nature of IN action dictates that its virgin matter can be highly concentrated or highly dispersed — or in any concentration and/or mixture in between.

In this case, the distribution in a highly dispersed pattern and the young age of the radioactive iron that is freshly made, relatively speaking, are clearly indicative of its origin via ongoing IN. A good analogy can be found in the tiny radioactive polonium halos found dispersed in granite. Each of the two elements could only have been made and highly dispersed in this manner via IN. There are, of course, other analogies described in my book, *THE SPACING OF PLANETS: The Solution to a 400-YearMystery* (1996).

The interpretations of the discoverers, Schewe and Stein, who believe that "all the heavy atoms of Earth have been processed through or created in supernovas long ago and far away" somehow have the sound of a fairy tale. And they do raise questions. If made in a distant supernova, why and how would the radioactive Fe-60 with a half-life of 1.5 million years manage to find our tiny planet in the unimaginable volume of space? Did it travel at nearly the speed of light straight to Earth? Can the similarly limited concentrations of polonium halos in granite be traced to an outer space source?

How can the following be explained except via the FLINE model of planetary origins?

1. The several iridium layers found on Earth.
2. The titanium layer found on the Moon.
3. The tellurium layer found on Venus.
4. The magnesium layer found on Neptune.

Did each (or any) of the above four types of layers selectively come from distant supernovas? Or did a different type of comet selectively crash on each one of them?

If so, why did the several iridium comets always crash only on Earth?

5. The abundant (virgin) metals observed at the impact site of fragment G of Comet SL9 on Jupiter.
6. The atmospheres and surface features of planets and moons, and why they differ in accord with size.

The answers are easy to understand when we realize that all are in situ ejecta products of IN; these things did not, and do not, come from outer space. If only scientists were given the opportunity to learn about the FLINE model, there would be no need to explain everything by simply saying that it came from outer space long ago and far away. Certainly, in this situation Schewe and Stein could have benefited immensely from this knowledge; e.g., they erroneously state that "...*the lack of terrestrial sources point to a relatively nearby and recent supernova as the origin*" *of the Fe-60*. Had they been aware that no active planet lacks a terrestrial source of virgin matter, and that all such planets are self-sustaining entities, there would have been no need to speculate about a supernova long ago and far away. The great potential for huge savings in research time and money could be realized if scientists would be willing to examine their findings in the perspective of this revolutionary FLINE model of planetary origins.

Thanks again for the information.

To: Grady Traylor, 245 Baywood Circle, LaGrange, GA 30240. 08-05-99

Thanks for your letter of August 4 in which you posed three questions. I prefer to answer them in writing (for the records) rather than calling, and in the chronological order submitted.

1. Concerning my letter of April 21, 1999 to Glenn Strait, Cooper, Hyne, Anderson, Whelan and you, there has been no response other than your letter today. But this is no surprise. Through the years of correspondence on the subjects of abiogenic origins of hydrocarbon fuels (1975-1999) and the FLINE model of planetary origins and five-stage evolution (1995-1999) (to scientists, editors, etc.), the rare responses have remained apprehensive, while they prefer not to get involved. I believe the reasons include: (1) the definitive threat these ideas pose to prevailing beliefs and their advocates; (2) the findings, based strictly on scientific facts minus speculations, leave no room for rebuttal — or else advocates of prevailing beliefs would have quickly and gladly pointed out a flaw of some type; (3) the two concepts put reviewers on the hot spot: they can find no fatal flaw, and they cannot afford to go against the established beliefs of their peers, which, in turn, would jeopardize their continued access to grant money and their favorable standing with their cohorts. Too much is at stake — which is understandable.

One example: my Fourth Law manuscript submitted to the Royal Astronomical Society, London has been held in review (or abeyance) now for 19 months. As the world's top scientific organization, they are truly on the hot spot: damned if they do, and damned if they don't, upset the apple cart of established beliefs about planetary origins.

2. Yes, I definitely am interested in sending information to any and all of the interested parties, if you get the addresses. The best and most comprehensive material can be found in my latest book, *THE SPACING OF PLANETS: The Solution to a 400-Year Mystery*. The contents are cumulative findings since the concept began in 1973 with the abiogenic origin of the hydrocarbon fuels. Herewith are two pieces of literature.

3. George Brawner was a resident and former coworker in the chemical lab during the early 1970s. We later corresponded about my first book, upon which occasion he wrote the statement about the immense ramifications of the new fuels concept. He also wrote that the discovery and proof are worthy of the Nobel prize. The last I heard, he had accepted a job in Chicago.

Thanks again for your interest and assistance. Maybe some day soon the truth will come out.

Powering the Next Century

The excellent article, *Powering the Next Century,* by Richard Stone and Phil Szuromi (*Science* 30 July 1999) contains a statement that warrants some comments. Speaking about the "unexpectedly" cheap oil prices in the United States, they erroneously state that such low prices were "impossible to foresee in the immediate aftermath of the [energy] crisis" of the 1970s.

The authors apparently remain unaware of my publications in the 1970s that accurately predicted the current price range of $1.00 per gallon of gasoline. These predictions and the logical reasons that made them possible can be found in *Fuels: A New Theory* (1975) and *Undermining the Energy Crisis* (1977, 1979). This revolutionary Energy Fuels Theory (EFT) has been continually updated and expanded in later books to include the mounting evidence that periodically comes with relevant discoveries.

Understanding the abiogenic origin of hydrocarbon fuels led inevitably to understanding how all planetary matter came into being in like manner via the processes of internal nucleosynthesis (IN) (the transformation of energy into matter). Inevitably, this led to a paramount issue: The mysterious spacing of the planets around our Sun. The solution (1980-1995) to the enigmatic Fourth Law of Planetary Motion that first eluded Kepler in 1595 details how the planets attained their orbital spacing around the Sun before beginning the five ongoing stages of evolution (E) from energy masses to the planets we see today, all in full accord with size. Combining the Four Laws of Planetary Motion (FL), internal nucleosynthesis (IN) and evolution (E) gives us the revolutionary FLINE model of origins of solar systems and the evolution of planets. These three chronological, inseparable and ongoing realities of the new model remain interactive until, one by one, they reach their ending. Thus, from energy mass to inactive sphere, each planet is a self-sustaining entity that creates its atomic compositional matter via internal nucleosynthesis throughout its first four stages of evolution.

Understanding the FLINE model enables one to make accurate predictions concerning planetary anomalies other than those of the hydrocarbon fuels. One example: Early in 1994 the prediction that the collisions of the 21 fragments of Comet SL9 on Jupiter would be the most powerful explosions ever witnessed in our Solar System was in stark contrast to the much weaker predictions of the best super-computers of the time. The laughter at this 'wild' prediction faded quickly after the powerful explosions at Jupiter's cloud-tops fulfilled all predictions of the FLINE model, while simultaneously casting doubt on the snowball concept (of comets) that served as the basic model for the computer predictions.

If the understanding of all planetary anomalies is to be accomplished, it must be done within the realm of the FLINE model of planetary origins and evolution. The confidence expressed in this conclusion is directly proportional to the quantity and quality of evidence for the conclusion. The new concept meets all validation criteria: (i) it makes successful predictions; (ii) it is based on observations that could, in principle, refute them, but have not; (iii) there is a comparable competing concept that is faring worse (one that fails to meet the validation criteria). An a priori test would present no problem with it.

R. Kerf's *USGS Optimistic on World Oil Prospects* (*Science* 14 July 2000) is the latest in a series of articles on the status of the world's petroleum reserves. As in every article since 1935, the reserves have been grossly underestimated, causing false and dire predictions about the exhaustion of world reserves. The cause of these inaccuracies resides in the erroneous beliefs of the biogenic origins of hydrocarbon fuels (gas, petroleum, coal). How these antiquated beliefs survive in the presence of today's voluminous and powerful substantive evidence against them remains a mystery, especially since the substantive evidence proving their abiogenic origins is even more voluminous and powerful. According to Nobel laureate, Irving Langmuir: "Pathological science is the science of things that aren't so."

Since the Copernicus/Galileo era, history has taught that the progress of science is linked inexorably to the openness of minds to new ideas. Otherwise, adherence to the status quo at all cost ultimately proves too high a price. Should science risk adherence to antiquated beliefs that can lead only to pathological science? Shouldn't we be willing at least to examine, with open minds, all evidence supporting the abiogenic origins of hydrocarbon fuels, the origins of planets other than via accretion, the origins and true nature of comets and asteroids, the sources of planetary waters (not from outer space!), the nature of planetary quakes, the most basic causes of lightning and other weather phenomena, the accelerating expansion of the Universe and even the Big Bang itself? Valid answers to these anomalies have their taproots deeply embedded in two inexorably linked concepts now known to, and accepted by, only a few scientists: the FLINE paradigm and its partner, the Little Bangs (LB).

The FLINE model was brought full circle via the solution to the proposed Fourth Law of Planetary Motion (1980-1995), explaining how the planets attained their orbital positions around the Sun. Other than offering definitive solutions to anomalies of the planets of our Solar System, the FLINE model details how each of the known (40+) extrasolar planets formed so close to its central star (not possible via accretion!). In close conjunction with the FLINE model, the LB model of universal origins explains the accelerating expansion of the Universe without the necessity of a cosmological constant.

The out-of-hand rejection of new ideas that challenge the status quo is perhaps the most effective way to impede scientific progress. The choice is clear: Either we adhere to outmoded beliefs that lead only to costly pathological science, or we can examine newer alternatives that offer definitive solutions to the mysteries of the origins of solar systems and the evolution of planets and moons by means of natural laws. Can science really afford the high price of ignoring the lessons of history?

A Secure Conclusion

This condensed book of the FLINE model presents a fundamentally sound, factual argument for a revolutionary version of the origin of our Solar System and the evolution of its planets and their planetary systems by means of natural laws. As with a giant jigsaw puzzle in which all pieces interlock precisely in place, it presents a beautiful and complete picture that offers a new perspective in which the enigmatic anomalies of the SS can be understood, and all relevant discoveries of the space probes interpreted more logically.

In the new perspective, scientists will be better able to understand the true cause of planetary quakes (e.g., earthquakes): explosions that cause land movements on those planets and moons during their active stages.

Subsequently, the source and cause of lightning will be understood. Recently, scientists have recognized that ground-to-cloud lightning is a reality – actually the norm, rather than a figment of the author's imagination. We can look to Earth's core for the supplies of electrons that make lightning possible. Inevitably, the understanding of all phenomena of Earth and the SS will follow.

But the path will not be easy. Obviously, the strong entrenchment of the prevailing Accretion Disk theory (the dust aggregation/planetesimals/accretion hypothesis) in the scientific literature presents one of history's most formidable barriers to change in direction of scientific thought. However, history teaches that more satisfactory concepts eventually do displace unsatisfactory concepts.

Inevitably, time will correctly judge the two viewpoints.

CHAPTER IV

MOONS, PLANETARY RINGS, COMETS AND ASTEROIDS

"We must forsake such beliefs that Earth is flat, that its core is iron or rock, that fuels were made from fossils, that energy is finite, that planets are created from dust and gases, and that continents still drift. Such concepts have too long misled scientists down blind alleys." Author (1980)

Origin of Our Moon

Our Moon is a gleaming silver globe that for centuries has inspired poets, artists, musicians and lovers: a huge, silent barren ball of rock that travels around the Earth in slightly less than one month's time. While it appears to be a pale silver in the daytime, it is actually a dark brown color. Brown is the color of cooled lava, pumice (volcanic glass) and igneous rocks that comprise the surface. Jagged, rocky mountains stretch across part of the Moon.

The silver sphere is 2,160 miles in diameter and one-eightieth the mass of Earth. Its surface gravity is only one-sixth as strong as Earth's surface gravity - too weak to retain an atmosphere. Any water or light gas vapors formed in the past would have evaporated immediately into space.

With its mountains, extinct volcanoes and moonquakes, it is easy to visualize the Moon as simply a smaller Earth without atmosphere or water. One can expect to find most of the same elements comprising both spheres. However, due to different internal conditions affected by size difference, the ratios of created and of retained elements should, and do, vary considerably.

Between the first landing on the Moon in 1969 and the last one in 1972, some 850 pounds of rock samples were brought back for analyses. The abundance of some of the main elements – silicon, magnesium, iron, manganese – in the two spheres matched. Refractory substances such as aluminum, calcium oxide, chromium oxide and titanium that are difficult to vaporize were quite different, as predicted by the FLINE concept. The Moon samples showed twice as much as Earth's contain. The biggest differences showed up in the more volatile substances, such as potassium and sodium: the Moon has much less than Earth. No trace of water was found on the Moon.

In 1982, Kirk Hansen of the University of Chicago suggested that the changing rotation rate of Earth determines the rate of tidal dissipation over geological time. His calculations argue favorably for simultaneous creation of Earth and Moon, thereby adding credibility to the contemporaneous geometric birth of the "double-planet" expounded in the FLINE model as early as 1980.

According to the FLINE (formerly the GBSST) concept, Earth and Moon were formed contemporaneously from the SEM as it sped beyond the Sun. Coupled into a giant dumbbell formation since then, the two masses trace an intertwined pattern in their revolutions around the Sun. The center point of the dumbbell's mass, rather than the center of the Earth, moves in a smooth ellipse around our Sun, causing each of the two masses to follow a serpentine path.

The results from the Clementine survey of the Moon yielded the first complete global portrait of Earth's orbiting partner. Among the findings in 1994:

- Volcanic activity as recent as a billion years ago.
- A wildly variable crust.
- The possibility of ice in the shadow of the South Pole.
- A crater large enough to span the USA from the East Coast to the Rocky Mountains.
- A fresh-looking crater that may have been made in the 12th century and recorded by monks.

Scientists had thought that nothing much had happened on the Moon in the last 3 billion years. The discoveries brought the realization that scientists do not understand the Moon as well as they thought they did.

Clementine's 71-day rendezvous with the Moon revealed a topography marked by steep peaks rising to 10 miles higher than the lowest valleys. The deepest valley is rimmed by the highest peaks: a vital clue to the early stages of formation when forces of isostacy and ejection (internal forces) worked together to push up the huge rim of pliable, amorphous material that solidified as mountains, while the remaining ejecta sank, filling the void below and forming the deep valley.

Evidence of recent lava flows in the Schröedinger basin suggests that the Moon was erupting with volcanoes perhaps 2 billion years after it was believed to have settled down.

Such volcanoes represent the second stage of mountain building in which ejecta must find a way out from beneath the now-solidified crust. Just as on Earth and other planets, virgin ejecta pushes through crustal places of least resistance to its pressurized flow to build the tall volcanic outlets.

This new evidence fits precisely into the FLINE concept in which these surface features can be traced to their source: internal nucleosynthesis. Additional evidence came in 1995 from McDonald Observatory in Fort Davis, Texas. Observations made during a total lunar eclipse in 1993 (released in 1995) show the faint glow of sodium gas some 9,000 miles in the atmosphere surrounding the Moon. Here the IN processes appear to be nearing their final stages, as attested by occasional small moonquakes and the weak out-gassing from craters.

The one big question left unanswered in many minds by the space probes is the Moon's origin. A growing consensus among astronomers favors the "giant impact" hypothesis in which the Moon may have gotten its start 4½ billion years ago when a planetary projectile about one-seventh Earth's mass collided with our mother planet. The energy of the collision crushed and vaporized major parts of the two masses, sending out a high-velocity jet of material at temperatures as high as 12,000°F. Within a few hours, some of it came back together far enough away from Earth to remain in orbit as our moon.

Proponents of this concept claim that it appears to explain the chemical findings from the Appollo mission; e.g., the moon rocks brought back lack water, sodium and other volatile materials – precisely the materials that would boil away in the rapid vaporization after impact. And some scientists believe the concept explains why veins of gold and platinum lie shallow enough in Earth's crust to be mined!

But not all scientists are satisfied with such a scenario. First, two of the above statements concerning sodium are contradictory. The 1993 observations of the faint glow of sodium gas surrounding the Moon up to an altitude of 9,000 miles directly contradict the 1980s impact scenario in which all sodium boiled away 4½ billion years ago. Further, the potassium and sodium reported in the early analyses of Moon rocks could not have survived such an impact.

And why would gold, platinum and other heavy metals be found only in a projectile and not in the larger mass of Earth? How were these materials made in the projectile? Would the missile have delivered all of Earth's other mined materials? Does this impact scenario apply to all other moons of all other planets? Aren't spherical moons all made in the same manner? Did the moons of Jupiter and Saturn result from projectiles bouncing off their clouds? Why didn't the exploding "asteroids" of Comet SL9 bounce off Jupiter to form more moons?

How dependable were the computer simulations that seemed to confirm the hypothesis? Were their results any more dependable than the erroneously predicted results of the powerful computer simulations of the collisions of Comet SL9 with Jupiter (as detailed later in this chapter)? Wouldn't the same type of simulation erroneously show that all moons of all planets were formed in similar impact scenarios? The fallacy of the impact hypothesis becomes obvious under the pointed finger of question.

As one scientist put it, "Books and articles supporting this giant impact hypothesis of lunar origin are more a testament of our ignorance than a statement of our knowledge."

As with the fallacious dinosaur theory, why is it necessary to look to outer space when better answers can be found more readily by looking to the nucleosynthesis processes within planets? In reality, all anomalies relevant to all moons and planets are explainable within the realm of the FLINE concept. For example, the Moon once generated a magnetic field, which may have been nearly twice as strong as the present-day magnetic field of Earth, as shown by Runcorn *et al.*, who used magnetized lava rock from the Moon as evidence.

The proof of the declining strength of the Moon's magnetic field is powerful evidence that such magnetism is (or was) of nuclear energy origin rather than of iron or rocky core origin. The large decline is attributed to the dwindling size of the core as its energy transforms into matter.

Other observations and studies of moons and planets have revealed magnetized lava rocks, the increasing of interior temperatures as a function of depth, the size and density of cores, the equatorial bulges, the convection of materials, the volcanic layers, the chemical composition of crusts, the ejecta and impact craters, the advanced stages of evolution, etc. All are symptomatic of nuclear cores.

In contrast to the impact concept, evidence supporting the FLINE version clearly shows that Earth and Moon formed contemporaneously and both have evolved via internal nucleosynthesis into the rocky fourth stage of evolution. The Moon is nearing the fifth and final stage: an inactive sphere.

In the final analysis, the Moon is nothing but a small and shining planet that evolved alongside Earth and in the same manner as Earth, partners from the beginning.

Shelley expressed his feelings beautifully when he wrote: "...that orbed maiden, with white fire laden, whom mortals call the moon."

To: David H. Levy, c/o PARADE Magazine, New York, N.Y. 11-06-00

While very interesting reading, your article, *Why We Have a Moon* (PARADE Magazine, Nov 5, 2000), presented the prevailing hypothesis of the manner in which our Moon was created: the Hartmann-Davis Giant Impact concept of its origin. In this incredulous version, "a small planet, with unimaginable energy and deafening noise, sideswiped Earth about 4.5 billion years ago, bounced off and, seconds later, tore right back into our planet with a super colossal force. That Mars-size world broke apart, and huge chunks of Earth's crust flew off into space. Two rings of debris, their particles much larger than those in the fine-grained rings of Saturn, grew and circled Earth. ...Pieces of the outer ring slowly gathered together around its largest chunks. In just a year, those pieces formed a large new world, a world we can still see. That world we call the Moon."

Any concept based on the Accretion hypothesis is already fatally flawed, with no chance whatsoever of being blessed with sound logic or factual evidence. This concept fails to meet any of the prerequisites for a valid theory as specified by true science. It is pure speculation, with no basis in fact; known facts about our Moon's origin present a powerful argument against this version. These facts, explained elsewhere in my books (1975-1996) and in other literature, are a vital part of the FLINE model of planetary origins and evolution. In brief, moons originate and evolve through the same five cycles of evolution as planets do; they are, in fact, simply smaller planets. Planets come in all sizes, from the smallest sphere to the largest giant exoplanet. The too-close proximity of each giant gaseous exoplanet to its central star is a powerful argument against the Accretion concept, but is strongly supportive of the FLINE model.

Many questions always will remain unanswered via the Giant Impact concept, but one question alone can illustrate the fallacy of this hypothesis: Considering all the moons of Jupiter, Saturn, and other planets, and realizing that all spherical moons were made in identical manner, did all these moons bounce off the thin cloud tops of their respective planet to become what they are today? If this sounds highly unlikely, then we must conclude that our Moon, or any other large moon, could not have been created via the giant impact-accretion hypothesis. Meanwhile, the FLINE model of creation of moons presents a valid scientific alternative soundly structured with irrefutable facts that leave no questions unanswerable or non-provable. An in-depth comparison of the two concepts readily exposes the ludicrous logic of the giant impact concept, which already is becoming a genuine embarrassment to science.

Herewith is a flyer describing my current book, *The Spacing of Planets: The Solution to a 400-Year Mystery*; it is online at iUniverse.com. My newest book is due online in January. The books contain answers to the many questions left unanswered by the Accretion hypothesis. Additionally, they detail the FLINE model's accurate predictions and interpretations of the results of the crashes (actually powerful explosions) of Comet SL9 on Jupiter. That event alone is conclusive evidence of the true nature of comets, which, in turn, is irrefutable evidence against the Giant Impact/Accretion hypothesis.

Recent Developments Concerning the Lunar Cataclysm Hypothesis.

In regard to *Support for the Lunar Cataclysm Hypothesis from Lunar Meteorite Impact Melt Ages* (*Science*, 1 December 2000), suffice it to say that the cogent evidence presented by B. Cohen *et al.* becomes exciting when more logically interpreted within the realm of the FLINE model. In reality, the data serve as powerful supportive evidence for Allan Mills' famous experiment that duplicated the surface of the Moon by forcing gases up through a layer of thick mud, thereby revealing the manner in which the crater-covered surface was formed via the ejecta concept of the FLINE model.

The event of a short, intense period of bombardment in the Earth-Moon system to account for these craters is a highly unlikely occurrence; this hypothesis isn't needed to explain the multitude of craters that can cover the entire surfaces of moons and planets. All that is needed is an understanding of Nature's principle of internal nucleosynthesis, the creative force without which evolution and ejecta craters would not be possible. Of course, this does not preclude the probability of occasional impact craters being formed when heavy ejecta matter fell back onto the surface, or when, on very rare occasions, an outer space object could have added to the sum total.

In the same issue of *Science* magazine, *Beating Up on a Young Earth, and Possibly Life* expounded on the lunar cataclysm. To quote: "Astronomers still don't have any good idea of the alleged lunar cataclysm's source. Simulations show that the gravity of Earth and other terrestrial planets would have cleared the inner solar system of threatening debris within a few hundred million years." According to the researchers, such "a cataclysm would require the breakup of a sphere larger than 945-kilometer Ceres, the largest Asteroid, and the chance of that happening any time in the past 4.5 billion years is nearly nil."

As a last resort, astronomers look to the outer reaches of the solar system, and speculations abound. For example, "Neptune and Uranus could have tossed icy debris, along with some asteroids, inward in sufficient quantities to resurface the Moon, give Mars a warm and wet early atmosphere, and sterilize Earth's surface with the heat of bombardment." One astronomer is toying with the idea that Uranus and Neptune started out between Jupiter and Saturn, where his simulations suggest they could have orbited for millions of years before flying out into the lingering debris beyond Saturn and triggering a late heavy bombardment! "That's my fairy tale," he correctly stated — a statement with which the new Fourth Law of Planetary Motion and the FLINE model fully agree.

The most amazing thing about this situation is that after 27 years of the evolving FLINE model's irrefutable facts and impeccable record of accurate predictions and five years after the discovery of the Fourth Law, many astronomers still prefer fairy-tale speculations over well-substantiated, scientifically valid arguments. Contrary to these *Science* articles, the origins and evolution of moons have been proven beyond any doubt by valid scientific methods. Even though the new model's powerful supportive evidence shouts loudly against the establishment's favored, but erroneous, hypotheses, it continues to fall on deaf ears. The blame must be placed squarely on the shoulders of advocates of the prevailing Big Bang/Accretion hypotheses.

If *Science* magazine continues to publish such speculative fairy tales in preference to substantiated facts, perhaps it should be renamed *Science Fiction*.

New Evidence Substantiating the FLINE Model of Creation of the Rings and Moons of Jupiter.

(Ref: THE SPACING OF PLANETS: The Solution to a 400-Year Mystery) (Scarborough, 1996).

On September 15, 1998, scientists announced to the news media that "the faint rings around Jupiter come from clouds of dust that are the result of cosmic debris battering Jupiter's small moons, according to data from the Galileo spacecraft." The rings — which are nearly invisible to even the best telescopes — clearly show their relation to the orbits of four small inner moons, the scientists said. According to Joseph Burns of Cornell University, "Pictures are the smoking gun that allows us to say this theory works."

Some questions become obvious: If it is true that planets accreted from dust, gases and planetesimals, one might question why the matter comprising the planetary rings remained in orbit rather than accreted into the greater mass? Why would this "cosmic debris" be battering Jupiter's small moons to create even more dust rather than accreting as larger bodies? Or is the condensation/accretion hypothesis of planetary creation erroneous? Can the relation of the rings to the orbits of the four smaller moons be interpreted as something more significant than a battering action that makes dust? What can be learned by comparing the rings of Jupiter to the rings of Saturn? A new perspective on the subject offers some answers.

Three fundamental and inseparable realities of nature comprise a revolutionary paradigm of planetary origins and evolution: the Four Laws of Planetary Motion (FL), internal nucleosynthesis (IN), and evolution (E). In any sphere, neither E nor IN can exist without the other. A critical connection exists between planetary evolution and the essential source of energy that drives planets through the five stages of evolution common to all such spheres — until each core is depleted of energy, leaving only an inactive (dead) sphere (e.g., Mercury) in orbit. This paradigm provides an understanding of the role of the inner core in various Earth processes, including the generation of the geomagnetic field and the thermal evolution of Earth. The IN/E concept leads inevitably to the mathematical solution to the pending Fourth Law of Planetary Motion (1980-1995) that first eluded Kepler in 1595. The Four Laws (the third fundamental reality) bring the FLINE paradigm full circle by explicitly revealing how nebulous planetary masses, including all exoplanets, attained their orbital spacing around a Sun.

Viewed in the perspective of the FLINE paradigm of origins of solar systems and the orbital spacing and subsequent evolution of planets, the data evolves into a much clearer picture of how planetary rings were and are created. First, we must recognize that all planets began as fiery masses of energy (small stars, the first stage of evolution) placed in orbit in full accord with the FL. During the first two stages of planetary evolution, these violent Sun-like masses ejected plasma of ionized gases, dust and solid matter that were rocketed into orbits as dust and small moons. In reality, the pictures identified as the "smoking gun" in support of the prevailing ring theory clearly reveal a pattern that fits precisely into the FLINE scenario. To corroborate this viewpoint, we can look to Saturn's older rings (now at, or near, peak accumulation) and observe their beautiful multi-patterned striations attained by the multiple ejection of selective matter created via internal nucleosynthesis and jettisoned into orbits at various times and at various distances around the mother planet.

All planets go through the early ejecta ring phase of evolution. The rings always end up circling the equator — just as the young and growing rings of Jupiter now appear to be in the process of accomplishing. Planetary size governs the rate of evolution through the five stages: the smaller the size, the sooner the evolution through each of the five stages. This tells us that Saturn is ahead of Jupiter in the forming of rings; thus its orbiting masses are a preview of the many more rings that will form eventually around Jupiter in similar manner. Of course, the early rings of our smaller Planet Earth and other small planets have long since disappeared onto planetary surfaces and into space.

At the latest count of astronomers in mid 2001, Saturn has edged into first place as the planet with the most known moons: 30, to Jupiter's 28, Uranus' 21, and Neptune's eight. The new survey turned up 12 new moons around the ringed planet to surpass Jupiter, whose number of moons has not been updated.

But not to worry; in the long run, Jupiter, due to its larger size and slower pace of evolution, will regain the lead in this interplanetary battle of numbers of ejecta moons.

The Far Side of the Moon

USA and Soviet lunar-orbiting craft have photographed portions of the far side of the Moon. Painstaking analysis of computer tapes indicates that some of the images provide an accurate, multi-wavelength portrait of parts of the far side. The images from Mariner 10 and Galileo, each depicting a different part of the far side, are helping astronomers decipher, in detail, the composition of the Moon's hidden half. Surface composition provides crucial clues to the nature of the volcanic eruptions and other upheavals that shape the surfaces of spheres.

Scientists have already mapped much of the chemical composition of the Moon's near side. Analyses of the lunar rock samples brought back by U.S. astronauts and unmanned Russian craft have proven extremely helpful.

Two types of terrain form the Moon's outer surface. The light-colored highlands represent the brighter, lower density minerals. In contrast, the dark plains are regions identified by Galileo and his Renaissance colleagues as maria, a Latin word meaning oceans. To them, the dark plains appeared to be bodies of water. Researchers generally agree that maria are the results of volcanic eruptions: lava flows from 4 billion to 2.5 billion years ago.

Images from the spacecraft confirm that the maria are much less abundant on the far side than on the near side of the Moon. This suggests earlier and then-milder volcanic activity on the far side – the latter due to the earlier, thicker crust there. But why would the crust be thicker on the far side? In addition, when did the maria form?

Evidence suggests that early volcanic activity that formed maria was extensive; some maria existed more than 4 billion years ago. According to the IN concept, the Moon's crust formed in the same manner as all other spheres of the SS, passing through the common stages of evolution while cooling from hot energy to cold matter.

During the transition stage, the Moon's thin crust struggled to survive and grow, eventually encapsulating the sphere with a rough, flexible blanket. Ever thickening, the soft crust became a thick, bubbling, mud-like caldron, stirred by escaping gases, matter and fireballs, all newly created within. These escape outlets hardened into circular patterns now erroneously identified as impact craters.

Meanwhile, the heavy atmosphere thinned and eventually vanished. Volcanic flows became common, covering many of the craters, even as others were being formed. Volcanic mountains and lava plains formed from huge outpourings of virgin materials that found the paths of least resistance to their upward pressures. Fireballs shot violently from within and fell back to the surface, sculpting the landscape with a multitude of ejecta and impact craters into the circular patterns and dark maria visible today.

Instruments that view the surface at several wavelengths can detect the dark, underlying maria, called crytomaria, mixed with the lighter-colored shroud of highland crust.

Patrick Moore, in his book *New Guide to the Moon* (1976 edition), presents a strong case for the formation of the craters of the Moon by internal forces rather than by impacting meteorites. Included is convincing evidence by Allan Mills, who successfully produced model craters strikingly similar to those of the Moon. He accomplished this proof by introducing a pressurized stream of gas below a particular material, which behaved as if liquid while the bed expanded. Bubbles appeared, venting through the muddy material to produce the crater-like appearance of the Moon.

Analyses of views of the far side show that the Moon's hidden half contains far fewer maria than the near side, and far lower concentrations of titanium oxide than do many maria on the Moon's near side. This evidence and the analysis of moonquakes and human-generated disturbances on the Moon have revealed that the lunar far side has a much thicker crust than on the near side. To the obvious question of why, the IN concept offers a reasonable answer.

As contemporary partners for some five billion years, Earth and Moon evolved with the near side of the Moon always facing Earth. During their evolution from energy masses to rocky spheres, the lunar far side remained exposed to the severe cold of space, while the near side basked in the heat from Earth's nuclear mass, a small secondary sun. These extreme conditions, along with the slightly larger centrifugal force acting on the far side, caused the crust there to evolve quicker and thicker than the much warmer near side evolved. Under these extremely different conditions, the two sides naturally evolved differently.

Noticeably more craters and titanium oxide on the near side are indicative of a later slower, longer stage of ejecta cratering. The difference in the amounts of titanium oxide is attributed to the fact that larger volumes of heavier elements such as titanium are produced in the later stages of evolution of any sphere. The steady increases in internal temperature and pressure – caused by the huge energy-to-matter expansion within – as the sphere becomes more tightly encapsulated accounts for the larger amounts of titanium oxide produced at a later time and over a longer period of time on the near side than on the far side of the Moon. The thinner crust adds another advantage to the near side by providing an easier upward route for the chemical ejecta.

One interesting observation here: The titanium produced on the Moon is analogous to the tellurium produced on Venus, the iridium produced on Earth, the magnesium produced on Neptune, and the abundant metals observed at the impact site of the G fragment of Coment SL9 on Jupiter — all are in situ ejecta products of nucleosynthesis.

In summary, the difference in the number of craters, the thickness of the two crusts and the amounts of titanium oxide are indicative of the differences in the rates of cool-down and solidification of the crust on each side of the Moon. These different rates, in turn, are functions of the outside temperatures to which the two sides were exposed and to the steady increases in internal temperature and pressure as encapsulation progresses as a function of time.

Other questions challenge the impact craters hypothesis. If by impact, why such a noticeable difference in the number of craters on the two sides of the Moon? In addition, why so many craters altogether on such a tiny target? Why are all craters lined up precisely with the center point of the spherical moon? Why don't any of the craters show a skewed configuration typical of a near-miss projectile hitting the surface at a sharp angle?

In conclusion, the evidence supporting the FLINE concept argues strongly against the giant impact hypothesis of the Moon's origin, while presenting a valid case for the origin of Earth and Moon as contemporary partners during their simultaneous evolution via nucleosynthesis from energy masses to rocky spheres.

Although beginning their evolution simultaneously, Earth and Moon have evolved at different rates because of the size factor. The Moon's energy core is nearing depletion, as attested by very weak signals from within. Approximately two-thirds of Earth's core has been depleted in nearly 5 billion years, leaving an estimated 2.5 billion years or more until it becomes inactive.

Meanwhile, its orbit will have moved approximately 20% closer to the Sun. During this long interval, Earth's climate will cycle through many changes while growing ever more barren. Our planet will follow Mercury, Moon, Pluto and Mars in becoming inactive spheres. Expectations that humankind can survive the gradual drift into these slow, drastic changes seem unreasonable.

How Planetary Rings Were Formed

In 1610, Galileo first observed Saturn's peculiar form – recognized several years later by Christian Huygens as rings. The next set of rings were detected in 1977 by James Elliot and his colleagues at

Cornell University when they noticed that the light from a star blinked several times just before Uranus occulted it. Voyager 1 sailed past Jupiter in 1979 and Saturn in 1980. Voyager 2 visited Jupiter, Saturn, and Uranus, finally sailing past Neptune in 1989. The spacecraft made the first detailed images to confirm the rings, and discovered rings around Jupiter and Neptune during the extensive SS tour.

How can the formation of these rings be explained? By observation, the rings in each case are located at the 'equator', the circumference of maximum centrifugal force. In the perspective of the IN concept, substances (gas, liquid, solid) were forcefully ejected from these spheres and collected at the circumferences of maximum force, and remained suspended in orbits around each active planet. The greater the escape velocity/mass ratio, the higher the orbit is of each substance. Regardless of any change in the tilt of the planetary axis, the rings will remain firmly in equatorial orbits while their velocities decrease over billions of years, and eventually will drop onto the planets to become small parts of the crust.

The same principles can be applied to other satellites of planets, whether Nature-made moons or man-made craft, provided the orbiting masses remain below a critical mass/velocity/gravity/distance relationship. Any mass exceeding this yet-to-be calculated relationship will gradually drift away from its central body. Uranus, Neptune and our Moon are examples of orbiting bodies known to be undergoing such drifting.

To some degree, when blasting satellites into orbit, mankind replicates Nature's procedure for orbiting the materials that formed planetary rings and some of the smaller moons. In this perspective, the discoveries of the rings of Jupiter, Uranus and Neptune were predicted by the IN concept before the author learned of their existence. By the same logic, it can be predicted that Pluto, because of its small size, is well beyond this ring stage of evolution.

Why rings on the larger outer planets and not on the smaller inner planets? The reasons can be found in their sizes, which control the relative rates of planetary evolution. In the gaseous stages of their transformations, all planets had rings of ejecta: gases, solids, and liquids that quickly froze solid. Like a flower in full bloom, Saturn is at the zenith of planetary rings, the goal that Jupiter will attain some time in the distant future – the stage through which all other planets of our SS already have progressed.

The smaller the planet, the earlier it passes through the ringed stage of its evolution, which ended when all materials had fallen out of their orbits in full accord with the natural laws of gravity. Nothing can stay in orbit forever. Uranus and Neptune are in the twilight time of passing out of their ringed stage, while Pluto should have passed out of it a long time ago.

Ejecta created via internal nucleosynthesis is the key phrase that firmly ties together all things: planetary rings, craters, volcanoes, planetquakes, moonquakes, fireballs, comets, meteorites, the moons of Mars (Phobos and Deimos), lightning, electromagnetism and all other phenomena involved in the evolution of physical worlds. All are interconnected; all can be traced to a common origin: a nuclear energy core in each sphere. In this perspective, all past and future discoveries of the space probes will be understood.

Jupiter's Rings: A Giant Leap Forward

Until early 1992, the prevailing Accretion Disk theory (the condensation/planetesimal/accretion hypothesis) taught that planetary rings consist of leftover debris that did not accrete. In February of that year, the news media announced that the Ulysses spacecraft boomeranged past Jupiter on Saturday, flying through intense radiation and an orbiting ring of volcanic debris on its way to study the Sun.

The underlined words above represent a giant leap forward in the right direction of thought in the scientific community concerning the origin of planetary rings. The nuclear energy core of Jupiter accounts for both the volcanic debris and the intense radiation experienced by Ulysses. Both findings bring current beliefs about planetary origins sharply into question. Since both findings are products of internal nucleosynthesis and were predicted by the IN concept, they are powerful evidence supporting the FLINE model that foretold these discoveries as far back as the mid-70s.

Comet Halley and Its Last Farewell

In March of 1986, five space probes encountered Halley's comet to photograph its nucleus and to analyze its composition during its most recent periodic visit to the Sun. The photographs made by Giotto, the closest probe, revealed some real surprises. The picture and data beamed to Earth showed a velvet black, peanut-shaped nucleus with a surface full of pits. According to the IN concept, the surface is the carbonized crust encapsulating its nuclear energy mass, and the surface pits are ejecta holes or friction marks formed in the past by powerful jets of forcibly ejected materials.

The Vega probe reported evidence of water, carbon dioxide and hydrocarbon molecules among its findings, all products of combustion. Temperatures within the comet's thin plasma coma were recorded to be well above the boiling point of water. Computer-enhanced color images of the comet were identical to those of the background stars, indicating similar brightness and high temperature gradients throughout each of the photographed bodies. These likenesses received little notice and less publicity, but remain a significant factor in deciphering the true nature of comets. In spite of this contradictory evidence, many scientists still cling to the belief that fiery comets are rocky snowballs!

Violent, bright jets of gas and fine matter spew profusely from vents (or craters) in the nucleus. The jets appear to be firing out through two or three vents comprising only a small fraction – about 10 percent – of the total surface, leaving some 90 percent of the surface inactive. This fact of no losses from the black surface is highly significant: It means that materials forming the tails and the coma (with a radius of 600,000 miles) could have come only from within the nucleus: the source of their creation.

Dimensions of the nucleus were estimated as 16 x 8 x 8 kilometers. Its surface area is approximately four times larger than expected, and thus, the low albedo of 2 to 4 percent means that Halley's comet is the darkest of all known bodies in the SS.

Scientists have difficulty in reaching firm conclusions about the nature of the observed jet-like features, and in interpreting new information to fit the concept of a dirty snowball of pristine materials that supposedly formed the SS. The basic reason for these difficulties lies in the fallacy of the antiquated accretion disk hypothesis that evolved from the Kant-LaPlace hypothesis of a gaseous dust-cloud origin of the SS – a tragic myth that continues to lead many scientists down blind alleys in which nothing seems to fit together.

The Giotto pictures confirm that inert shells do develop on commentary surfaces, much like crust forms on planets – an exciting quantum leap. The next step is for skeptics to realize this startling fact. From there, it will be much simpler to understand how moons, planets and their systems were made in compliance with the laws of thermodynamics.

Both Giotto and Vega spacecraft carried aboard them sensitive instruments to identify gaseous molecules in Comet Halley's coma. They found a veritable stew of molecular fragments known as free radicals, confirming conclusions made previously from telescopic observations. Some 75-80 percent of the radicals pertained to water, a ratio common to Earth's water-to-land ratio produced via nucleosynthesis!

The remaining free radicals suggest that carbon dioxide, ammonia, methane and hydrogen cyanide would be the end products if these highly reactive radicals could get close enough together to interact and combine. However, most remain too far apart to combine into these compounds. It is no coincidence that the same compositional radicals are found in stars, and the same end products they form are found on all planets and moons of the SS. All were, and are, produced via nucleosynthesis within the fiery core of each evolving mass.

Other than these gases, and not surprisingly, a major discovery was the existence of tiny, solid CHON particles. CHON is an acronym for the chemical symbols of the compositional elements: carbon, hydrogen, oxygen and nitrogen. While these compounds are not biological in origin, they are indicative of compounds that might appear as black soot made via nucleosynthesis within the fiery comet. (Note the similarity with the nucleosynthesis of hydrocarbon fuels in Chapter III). This accounts for the velvet-black surface of the nucleus, perhaps a carbonized insulator against the tremendous heat of its core, and the near-absolute cold of space — analogous to the formation of crust on all planets and moons.

Large quantities of very fine particles were gathered by the dust collectors aboard Giotto and Vega. Chemically, these motes consist of many of the same elements comprising terrestrial rocks: iron, magnesium, silicon, oxygen, etc. The presence of these virgin elements, created via nucleosynthesis within the comet, further validates the FLINE concept and its predictions.

Since 1987, observers have detected methanol in seven comets. It now appears that this compound ranks third in abundance among commentary ejecta, behind water and carbon monoxide. The methanol can amount to 5 percent or more (relative to water) of a comet's volatile matter.

The results from Halley's Comet reveal its true nature. Rather than 'dirty snowballs', comets are what they seem: nuclear fireballs ejected from larger nuclear masses (much like flares ejected from the Sun), and powered in their orbits like nuclear-fueled jet engines. This mental picture was brought sharply into focus on the TV program, *Science Frontiers,* on June 7, 1994. Some eight years after the big event, the public was privileged to watch video close-ups of Halley's Comet zipping through the sky, looking precisely like a jet-propelled rocket in stabilized flight (see illustration).

Records show that the first recorded sighting of Halley's Comet was 240 B.C. It has returned periodically every 76^+ years, each time diminished in size as energy transformed continually into ejected ionized matter. The original gigantic comet must have been a terrifying sight in its early years. In 1456, it frightened everyone in Europe so much that Christian churches added a prayer to be saved from "the Devil, the Turk and the Comet". As late as 1910, it still furnished a good, but smaller show. But in 1986, it was a disappointment to observers, who had to look closely to find it.

Then in February, 1991, when Halley's comet was 1.3 billion miles from the Sun, about midway between Saturn and Uranus, it suddenly expanded to 180,000 miles across and shone to more than 1,000 times brighter than normal. The event was startling and unique, totally unexpected so far from the Sun, and remains unexplainable in the perspective of the "dirty snowball' concept.

However, viewed in the FLINE concept, the spectacular event is understandable: the comet simply exploded in its death throes. It was the last sighting of the famous comet – the last farewell – never to return on another mission to brighten Earth's skies. Other known comets have had catastrophic endings; some have split into two or more pieces, others have crashed into larger spheres, including the Sun. The Biela comet broke in two in 1846 and has, since, disappeared. In reality, each fiery comet exists for a relatively short time, usually measured in centuries, before becoming extinct. Rather than primordial remnants (dirty snowballs) left over from the formation of planets, they are more like a fiery species that refuses to become extinct.

Comets and Asteroids: Keys to Planetary Origins?

The ever-present dangers of comets and asteroids crashing into Planet Earth and wreaking tremendous havoc worldwide have caught the attention of the public. The extinction of the dinosaurs is often used as an example of what can happen when a large object from outer space creates a nuclear winter — although this scenario remains in dispute. Many scientists believe that volcanism was the culprit — and there is much corroborative evidence favoring this concept. To understand Earth's mysteries — past, present and future — we must understand the origins of solar systems and the evolution of planets. But to fully comprehend planetary origins, we must know, beyond doubt, the true nature of comets and asteroids and their sources.

In spite of much evidence to the contrary, many scientists cling to the belief that comets are gigantic masses of ice, snow and dirt from outer space far beyond our Solar System. This icy-conglomerate hypothesis of comets was brought forward in 1950 by Fred Whipple, and modified somewhat by Gerald Kuiper in 1951. In 1987, William Hartmann and his colleagues defined a comet as "a body formed in the outer Solar System containing volatiles in the form of ices and capable of developing a coma if its orbit brings it close enough to the Sun." The accretion concept of the origin of our solar system and the evolution of planets is based on the belief that comets, asteroids, dust and gases are pristine materials that came together to form these systems. Keep in mind that this hypothesis is structured with speculation, with no basis in fact. But it soon may be facing a serious challenge to its validity. Let's see why.

When recent close-up photos of comets are examined carefully, they present some startling revelations. Always conspicuous is a brilliant white mass that looks precisely like a white-hot fireball blasting from the rear of the relatively tiny, pitch-black nucleus (Figure 7), followed by a number of jet streams formed into a fan-shaped pattern of material (Figure 9, 10). Ancient drawings of comets and recent photos reveal separate jet streams usually numbering between two and ten or more. This pattern of distribution gives an indelible impression of jet streams forcibly ejected at various angles (against the solar wind) from several portholes, or craters, in the nucleus. These comet tails, millions of miles long, consist of ionized gases and dust comprised of a variety of chemical elements like those observed on the surfaces of the Sun and Earth. An ultraviolet image reveals that the hydrogen coma of comet Kohoutek had a diameter of more than 3,000,000 miles, nearly four times the diameter of our Sun. Such a large spherical coma is typical of all comets; for equal size comets, the younger the mass, the larger the coma. The fact that this thin veil of hydrogen extends so far in front of the tiny nucleus signifies a powerful and constant driving force capable of putting, and sustaining, it there against the resistance of the solar wind.

Photos: Max-Planck-Institute Fuer Aeronomie, Lindau/Harz, FRG; taken by the Halley Multicolour Camera on board ESA's tiotto spacecraft. Interpretation is by the author.
Figure 7

These exciting photos of Halley's Comet clearly reveal its fiery nature. Note the brilliant white jets firing from the rear of the fireball's irregular crust, locking the nucleus in a precise flight position in each picture — much like a jet-powered rocket When the core's nuclear energy finally is exhausted via conversion into matter, the nucleus may remain as a pitted, ejecta-cratered chunk resembling the two moons of Mars: Phobos and Deimos.

Note: Time elapse for photos was approximately one hour.

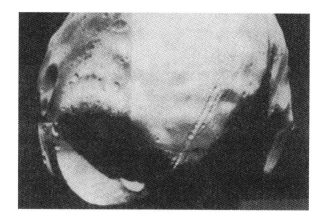

Phobos, a burned-out cometary nucleus.
Figure 8

The article entitled *Rocky Relics* (*Science News*, 5 Feb 1994) discusses near-Earth asteroids (NEAs) and the possible relationship of comets, asteroids and meteorites. It states that in 1992, the asteroid known as 1979VA was identified as the same object that had been identified in 1949 as a comet (Wilson-Harrington) with a "faint, but definite tail...Researchers now say that the comet and the asteroid are one and the same." This finding adds much confirmatory evidence and credibility to the same conclusion reached by the IN concept in the 1970s. It represents another giant step in the right direction.

Another quote from the same article states that "several bodies classified as asteroids may once have been comets. One candidate is the asteroid 3200 Phaethon. It follows the path of small bodies that produce the Geminid meteor shower, the flashes of light visible each December. Meteor showers are typically associated with comets. Tracks of dust expelled by active comets as they pass near Earth's orbit produce the flashes as the dust burns up in the atmosphere. Following this line of reasoning, Phaethon may be a comet masquerading in its old age as an asteroid."

In his *Principles of Philosophy*, Rene Descartes (1596-1650), the inventor of the analytic geometry used in the recent solution to the Fourth Law of Planetary Motion, defined Earth's interior as being "Sun-like" — a brilliant insight, especially since internal nucleosynthesis was unknown in his time. Without any evidence, he accurately defined the true nature of all active planetary cores. Planets and moons can still show clear evidence of being cometary masses, and recently have been recognized as such by scientists — although they stop short of the Sun-like version. In 1988 scientists detected traces of sodium and potassium as components of the Moon's extremely thin atmosphere. Three years later, researchers discovered that the atmosphere stretches out into a long tail about 21,000 kilometers in length and pointing away from the Sun. A corona of atmospheric atoms was found to extend 7,000 kilometers above the lunar surface —much like the coma of a comet. Their observations suggest that the Moon's atmosphere resembles that of a comet with an extremely tenuous tail; they also serve as powerful corroborative evidence of the evolutionary concept of planetary spheres from their original energy masses.

Some astronomers concede that the Moon does have the appearance of a comet. This is a giant step toward confirming my 1975 realization of the manner in which all solar system spheres were, and are, made via internal nucleosynthesis within their energy cores. Each sphere is a self-sustaining entity that does not require help from outer space to account for its ever-changing surface features and species.

These facts led to the solution to the proposed Fourth Law of Planetary Motion (1980-1995) that first eluded Kepler in 1595, and now explains how the nebulous planetary masses attained their orbital spacing around the Sun. In like manner, the Four Laws (FL), along with the corroborative evidence presented herein, also explain the origins and evolution of the exoplanets. All active planets continue as self-sustaining entities until their energy cores are depleted (all energy transformed into atomic matter), at which point they enter the fifth and final stage as inactive, or dead, spheres.

Valid arguments can be deduced from the evidence above to identify and confirm the sources of comets. Their ongoing birth-and-death cycle points clearly to the connection between the craters (primarily ejecta type, but some impact) of planets and moons as the byproducts of periodic ejections of these smaller cometary masses from larger energy masses. But our Sun will always be the main source of the fireballs of energy we call comets.

Many comets survive their cycles without exploding or falling to fiery deaths into larger masses. What's left after the cometary burn-outs are recognized as asteroids of a volcanic nature and appearance. Close examinations of the ejecta craters of the two moons of Mars (Phobos and Deimos, Figure 8) confirm their volcanic and ejecta nature beyond doubt. But what is, or was, the source of other types of asteroids?

In 1802, Heinrich Olbers, Germany's leading expert on comets, recognized the asteroids orbiting in the fifth orbit from the Sun (between Mars and Jupiter) as fragments of a planet that has disintegrated, exploded. This accurate interpretation accounts for the more than 30,000 asteroid fragments, most of which would have begun their existence as small, fiery comets and smaller, shattered fragments we identify as meteorites: lumps of stone or metal that survived fiery passage through Earth's atmosphere. In all probability, the surviving asteroid fragments represent only a fraction of the original number of shattered pieces in orbit. Evidence of current orbital patterns indicates the probability that three small planets existed in three distinct orbits before their disintegration into fragments.

The Phi geometry in Chapter 1 shows that the Asteroids planet might have gone into orbit as a single mass at 4.24 AU. Later, due to its vulnerable position between the Sun and the giant Jupiter, it broke free into three fiery masses that settled near their current orbits (2.2 AU, 3.2 AU, 3.9 AU) before the explosive disintegration of each of the three potential planets resulted in the three wide belts of orbiting asteroids.

In 1997, some 500 pictures of the heavily cratered asteroid, 431 Mathilde, were recorded by the NEAR spacecraft as it passed within 1200 kilometers (km) of the 50-km rock. Of the numerous craters on its surface, five measured more than 20 km in diameter. Casual examinations of the photos and these figures raise some key questions: How could this asteroid exist after its alleged collisions with so many huge projectiles, especially if comprised of accreted "rubble loosely bound by gravity?" Why the low density and dark color of the rock? Why does it rotate so slowly (once every seven days) after such alleged battering by the projectiles? In reality, no logical answers seem possible in the current perspective of the accretion or leftover debris hypotheses of the origin and nature of asteroids.

Research provides data, but data alone don't provide answers; it remains useless (and often harmful) until accurately interpreted. When these findings are analyzed in the perspective given to science by Olbers in 1802, the answers to these questions begin to make sense. As is well known, hot masses (e.g., lava balls) have quick-cooling surfaces: a dark, encapsulating coat that grows ever thicker as a function of time. During progressive stages of cool-down of asteroids like Mathilde in the super-cold vacuum of space, out-gassing and dramatic shrinkage of solid matter undergoing entrapment result in the collapse of huge areas looking much like smooth, shallow bowls. The bowl-shaped areas usually are accompanied by smaller porthole-like craters scattered about, small craters that had once served as outlets for the hot, pressurized gases and other material. Mathilde-type asteroids could have evolved only from the scattered magma from their exploded planets — which accounts for their lower densities and the darker color and surface features revealed so clearly in the photographs.

By similar logic, we can reason that stony asteroids are remnants of the original surfaces of the planets, while the nickel-iron masses came from within, or near, the planetary energy cores, either before or during the dramatic breakup of the three planets. Thus, the creation of different types of asteroids can be accounted for by the remnant masses undergoing rapid cooling in the super-cold vacuum and weightless conditions of space. Whoever determines the correct age of any stony or metal asteroid should establish precisely how long ago the three planets exploded into the tens of thousands, or more, asteroids.

While arguing strongly against the current beliefs of many scientists that planets formed from accreted rocky snowballs or icy mudballs, these findings clearly show the close relationship among fiery comets, cratered asteroids and moons, planets, and our Sun, thereby adding more credibility to the FLINE concept of our planetary origins.

What happens to comets that do not meet catastrophic deaths before all their internal energies become depleted? We can look to Mars for two good examples of burned-out commentary masses: its two moons, Phobos and Deimos. Both of the small, irregular bodies exhibit typical characteristics of burnt-out comets: ejecta craters and parallel striations or grooves. The densities of both moons, like the volcanic rocks of Earth, are much lower than that of Mars. From certain angles, Phobos even bears a striking resemblance to the rear end of a jet engine.

Structured with the brilliant works of Chladni, Kepler, Olbers, Descartes, Mills *et al.*, and corroborated by recent space discoveries, the FLINE paradigm offers a valid alternative to current beliefs about planetary origins and evolution — a paradigm made possible via evolving comprehension of the true nature, history, and relationship of comets and asteroids. These often terrifying missiles are indeed crucial keys to planetary origins and evolution.

Great Comet of 1861
Note the 8-tail ejecta from 8 crater ports.
Figure 9

Morehouse's Comet
Note the multi-port ejecta pattern.
Figure 10

Galileo's view of the cratered asteroid 243 Ida.
Figure 11

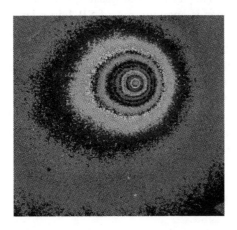

Color-enhanced **Halley's** Comet showing concentric enveloping matter indicating brightnes and temperature differentials. Nuclear stars in the background had identical colors and patterns indicative of nuclear compositions.

Figure 12

Alexander A. Scarborough

Hale-Bopp: A Great Comet

On July 23, 1995, two amateur astronomers independently discovered their namesake comet, Hale-Bopp, blazing away beyond the orbit of Jupiter 25,000 times brighter than Comet Halley did when it was at the same distance from the Sun. As it swung ever closer, the new Hale-Bopp mass put on a spectacular show in March and April 1997, a spectacle reminiscent of the early appearances of Comet Halley between 12 B.C. and 1682 (and to an even lesser degree until 1910) when its brilliance caused a sensation in the sky every 76^+ years.

How Comets Stay Active in Near Absolute Zero Cold

Assuming comets are ice-rock masses, temperatures in the far reaches of the Solar System are too low for compounds to sublimate from their hard-frozen surfaces. However, astronomers have discovered that some comets, including Halley, experience brief flare-ups of activity, and others such as Schwassmann-Wachmann 1 (SW1) retain their persistent coma. The claim of some astronomers that sublimation of specific gases account for these phenomena in near absolute zero conditions seems unrealistic.

In the cold, dim environment of the outer SS, specific radio emissions characteristic of any sublimated gas molecules would be so faint as to be undetectable. In addition, a coma would be non-existent, leaving the comet's small nucleus invisible even through powerful telescopes.

But the nucleosynthesis (fireball) version circumvents these problems. In it, the comet continuously produces trails of gas and ultra-fine dust; however, with increasing distances from the observer, these tails soon become invisible, while still sending out detectable radio emissions. For example, faint detection of CO has enabled astronomers to estimate that the very distant SW1 is emitting CO at the rate of about 2 tons per second. "This is remarkable," stated one researcher, "because it is comparable to the out-gassing rates of comets" [in the inner SS]. The style of out-gassing is comparable, too. A slight Doppler shift in the recorded line implies that the gas is spewing out of SW1 <u>in a jet</u>, supposedly like those produced by water vapor in comets closer to the Sun. The expulsion of other molecules (gases) such as nitrogen cannot be ruled out: astronomers have no way to detect them at such distances at the present time.

Without the continuous activity of nucleosynthesis in these very distant comets, the existence of coma and production of gases and dust would not be possible. Perhaps some day soon, improved detection methods will enable astronomers to detect and verify that the other gases and matter do continue spewing in a jet-like fashion from comets in the deep cold dimness of the outer SS – just as predicted in the 1970s by the IN theory. This will be another giant step favoring the FLINE model of planetary origins and evolution over the antiquated accretion concept.

More on Comet SL9's Collision With Jupiter: Some Stunning Results

"What people don't know isn't nearly as aggravating as what they do know that isn't true."
Mark Twain

In March of 1993, a fragmented comet was discovered moving away from Jupiter. Because of their bright, in-line formation, the Shoemaker-Levy 9 (SL9) masses were immediately dubbed "a string of pearls". Scientists designated the glowing fragments of SL9 by the letters A through W.

The following anomalies will be viewed first in the perspective of current beliefs about our planetary origins and then in the perspective of the FLINE concept. The reader will note a major difference in the understanding of what really happened in each of the mysterious events.

On July 16, 1993 the segmented comet reached its greatest distance from Jupiter before heading back toward the giant planet. Inexorably moving in a lengthening train toward their target, some of the fragments appeared to be breaking up, some growing dimmer. The small fragments, J and M, disappeared from view. Exactly one year later the bombardment of Jupiter began.

Jupiter's radio noise sounds like regular background white noise, except that it ebbs and flows in what has been described as the sound of the surf along a shore. Forty-three minutes after the first impact occurred, (the time it takes for radio signals to travel from Jupiter) the Jovian sound changed to a "rat-a-tat" static that lasted for nearly seven minutes.

A quote from the *Astronomy* magazine reads: "Radio astronomers expected radio emissions at high frequencies to drop when the SL9 comet crashed into Jupiter in July, 1994. Instead ... the JPL in Pasadena, California, found emissions increased 20 to 30 percent at a frequency of 2.3 gigahertz. Astronomers thought dust kicked off the comet fragments would absorb the electrons spiraling around Jupiter's magnetic field that causes the radio waves. But apparently the dust somehow created extra electrons that intensified the radiation through an, as yet, unknown mechanism."

In the new perspective, the string of pearls glowed so brightly because they were exactly what they seemed: freshly exposed nuclear fragments of a disintegrated comet. Any other masses of their small sizes would not have glowed like a string of pearls for such easy visibility. The two smaller masses, J and M, disappeared, due perhaps to either dissipation, or an over-coating of carbon.

The unexpected increase, rather than the expected decrease in electrons, is precisely what happens when a huge amount of energy is instantaneously added to a system. The "extra electrons" came from the exploding nuclear fragment. Any dust, if present, would have played an insignificant role. Only a nuclear explosion could have caused the 7-minute "rat-a-tat" static.

Beginning July 16, astronomers were stunned by the tremendous explosion caused by the fragments of Comet SL9 crashing against the thin cloud tops of Jupiter. Each powerful collision far exceeded supercomputer calculations. Comparisons of expectations with actual results were startling, even shocking to scientists. The *Astronomy* magazine (Nov 1994) published a well-written article evoking full agreement that pre-crash expectations were among the casualties. The spectacular explosions were indeed vastly more brilliant and powerful than predicted by supercomputers.

The information in this paragraph is from the record of predictions made by researchers (Eos, 5 Jul 1994). "Prior to the big event, three-dimensional simulations of the impacts on Jupiter were performed on the 1840-node, Ontel Paragon, the world's most powerful parallel computer, located at Sandia National Laboratories, Albuquerque, NM. The key results of the fireball simulation for a <u>3-kilometer</u> fragment impact were moderate, but seemed sufficient for the event. At 70 seconds after impact, a spherical shock wave has reached a diameter of 700 km and an altitude of 900 km above the clouds. Hidden within the spherical shock wave is the fireball itself, which is a rapidly rising cloud of cometary debris and Jovian atmosphere at a very high temperature (1700 K). Its apparent magnitude of 2 is about one-fiftieth as bright as Jupiter itself. The visible fireballs will be orange, fading to red, and most of the energy will be radiated in the near infrared."

Adding to the predictions, an article in *Science News* (9 Jul 1994), headlined *The 200,000-Megaton Meeting,* stated that each chunk is packing a punch that may exceed 200,000 megatons of TNT, and the expanding fireball may be as wide as 100 km when it emerges through the cloud tops within a minute after collision.

These recorded predictions later could be classified as puny and far off target when compared with the actual show put on by the impacts. The collisions were stunningly spectacular, the greatest ever witnessed in recorded history. Instead of small, orange flashes one fiftieth as bright as Jupiter, the huge, white-hot explosive flashes were up to fifty times brighter than Jupiter. Expended energy estimates shot upwards to 250 million megatons of TNT, then to <u>600 million megatons, creating temperatures of more than 30,000°C.</u>

Fragment G, alone, propelled a fireball thousands of kilometers above Jupiter's stratosphere, and is thought to have yielded at least 6 million megatons of energy. (A megaton is the equivalent of a million tons of TNT). To expend an equal amount of energy, a Hiroshima-type bomb would have to be detonated every second for ten years.

Infrared radiation (heat) from the explosion was so great that detectors at Keck Observatory were overwhelmed, or saturated. One collision was estimated to be the equivalent of 500 million atomic bombs. Astronomers reported seeing fireballs (from two of the smaller fragments) that erupted up to 1000 km above the cloud tops. After fragment C hit, astronomers at Keck Telescope in Hawaii took infrared photos. The views show two glowing ovals, each about the diameter of Earth, left by fragments A and C.

Other computers proved equally wrong. First, the fragments exploded at the cloud tops (a crucial clue to their true nature) rather than penetrating deep enough to dredge up Jovian material, disappointing many who hoped the impacts would reach Jupiter's water-rich lower atmosphere. Researchers reported that infrared spectra from Galileo suggest that comet fragment G exploded high in Jupiter's atmosphere, never penetrating the uppermost cloud layer. Analysis of much weaker spectra from the R fragment supports this conclusion. Another team of researchers deduced, from absorption lines in their spectra, that relatively little atmosphere resided above the G fireball, indicating that the fragment had not plowed into Jupiter's thin cloud tops.

The anticipated problem of separately identifying Jupiter's water and the supposedly icy water of the comet never materialized. After the largest collision by fragment G, astronomers were puzzled by their failure to find the chemical signature of water in the clouds created by the comet's impacts on Jupiter. "It's puzzling, but we will continue to look for water," stated one astronomer.

Scientists studying one of the early impacts found evidence of two chemical species never seen before on Jupiter: methylene (a carbon compound) and the hydroxyl ion. Both species would seem to require high heat to be formed.

Together, these findings raise serious doubts about prevailing beliefs concerning the nature of these brilliant streamers across the sky. If they really are 'dirty snowballs', there should have been obvious signs of copious amounts of water released in the collisions. And how will scientists explain the gigantic discrepancies between the supercomputers' puny predictions and the unexpected stunning immensities of the explosive collisions?

There can be only one reason for these huge discrepancies. Computers will put out correct information only when correct information is fed into them. Had they been fed information based on the correct assumption that the fragments were of a nuclear energy nature rather than an ice-rock composition, the computer would have been precisely in line with the stunning results.

The use of this correct assumption in a science contest of write-in predictions enabled me to correctly predict that mankind would witness the greatest explosions ever recorded within our Solar System. That prediction became a laughing matter among the experts, who, even today, still cling to the concept of rocky-snowball comets simply because it plays a crucial supporting role in the current Big Bang/Accretion theories about our planetary origins.

Further, in 1794, the true nature of comets was identified and published by the father of acoustic science, Ernst Chladni, who correctly argued for the cosmic origin of cometary fireballs and the generic connection between fireballs, meteorites, and meteors. The first proof came in January 1866 via spectroscopic observations of Comet Tempel-Tuttle that clearly showed it was shining by both reflected sunlight and glowing gas, and by its own light from the bright nucleus region that was too small to be seen through the telescope of the observer, William Huggins. And thermal radiation from the region of a nucleus was observed first by Carl Lampland as early as December 1927.

Just recently, comets were found to have a small magnetic field and to give off X-rays: all these properties are inherent in nuclear masses; all were anticipated by the fireball concept.

New Principles of Origins and Evolution
Revolutionary Paradigms of Beauty, Power and Precision

The obvious answers to all questions concerning these memorable collisions lead inexorably to the conclusion that the spectacular explosions were indeed of a nuclear energy nature. This rare event should be the final nail in the coffin of the 'dirty snowball' hypothesis, a false speculation that too long has misled scientists down blind alleys.

In reality, comets flashing through outer space are precisely what they appear to be: masses of hot nuclear energy that should not be confused with inactive asteroids composed of rocky matter that does not burn brightly in their flights through the vacuum of space, and thus remains invisible to the naked eye.

The evidence favoring the ancient belief that comets are of a fiery nature is overwhelmingly persuasive. But in spite of this powerful evidence, astronomers are not ready to change their minds. But there is hope: NASA's Deep Impact Mission is scheduled to launch a robotic spacecraft in January, 2004 to arrive at Comet Tempel 1 in July, 2005. Once there, it will launch a heavy projectile, allegedly, to blast a hole in the celestial body some seven stories deep and about the size of a football field in an attempt to peer beneath its surface. Researchers hope the impact will yield dramatic scientific breakthroughs — and it will! Let's hope that no one is within 90,000 miles of what will be the most powerful explosion ever triggered by mankind. This event also will trigger the falling domino effect on scientific beliefs about our planetary origins, including the four major myths: the planetary accretion, fossil fuels, BB and snowball hypotheses. The 21st century will be an exciting time of change in the direction of scientific thought about our planetary origins, thanks to Einstein's ubiquitous $E=mc^2$ that accounts for all the majestic beauty and awesome power of energy masses: from the smallest comet to the largest star.

This new knowledge about the true nature of comets should dispel the notion of 'dirty snowballs' from which our planets supposedly came into being. If not 'dirty snowballs', then the building materials for planets, moons and systems universal must be the one thing that is distributed throughout the Universe, the one thing that comprises all atoms and molecules of matter, the one thing that furnishes a solid basis for understanding all anomalies of our SS, the one thing capable of explaining all the stunning results of SL9 impacts on Jupiter: nuclear energy. The picture below clearly shows the white-hot nuclear power of nine Comet SL9 fragments on their way to colliding with Jupiter.

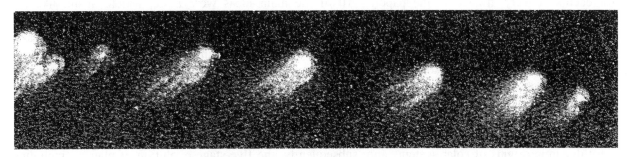

Comet SL9: Jupiter's fiery "string of pearls."
Figure 13

From: Amanda McCaig, Editorial Assistant, *Astronomy & Geophysics*, **the Journal of the Royal Astronomical Society, London.**
To: Alexander A. Scarborough, LaGrange, Georgia 30241   2001-03-28
Re: On the spacing of planets: The proposed fourth law of planetary motion- our ref. qy 016

We apologise for all the delays in handling your paper, which we accept fell below the standard expected of any academic publication. However, we have had enormous difficulty in finding anyone willing to referee your paper. Both the eventual referees agree that your paper is not suitable for publication in A&G.

To be specific, we must reject your paper on the following grounds:

1. The author does not cite any modern scientific work other than his own ideas.
2. It is not clear what deficiencies in currently accepted theories the author wishes to address.
3. The author's ideas are presented as assertions of fact without logical proof or any supportive observations.

Any paper that does not meet these academic conventions must be regarded as fundamentally flawed. We cannot therefore accept this paper for publication in A&G.

Please note the Editor's new address is:

Dr. Sue Bowler, Editor, Department of Physics & Astronomy, University of Leeds, Leeds LS2 9JT
Department of Earth Sciences, The University of Leeds, Leeds LS2 9JT

To: Amanda McCaig, Department of Earth Sciences, University of Leeds, UK. April 10, 2001
From: Alexander A. Scarborough
Re: qy016 On the Spacing of Planets: The Proposed Fourth Law of Planetary Motion.

Thank you for the valuable feedback on my paper submitted in January, 1998 to the Royal Astronomical Society for peer review and publication. I had predicted openly that you would have the enormous difficulty, as you stated, in finding anyone willing to referee the paper; the reason may well be that it is too revolutionary, somewhat analogous to the situation when Copernicus first placed the Sun at the center of the Solar System. The Fourth Law that first eluded Kepler in 1595 will have an equal impact on changing the way scientists view the origins of solar systems and the evolution of planets and moons. I especially appreciate your giving me the three specific reasons for rejection of the paper. Please allow me to offer some thoughts on each one in numerical order:

1. During the 23 years of carefully separating myths from facts and finding the solution to the elusive Fourth Law, I was equally dutiful in rendering proper credits in the body text to the person(s) responsible for each discovery utilized in the paper. All of the main ideas were mine, which, again, like the heliocentric idea of Copernicus, needed no contributions from the modern beliefs it opposes and exposes. Keep in mind that this is a revolutionary alternative to accepted theories, and like the Copernican idea, it is a definitive concept that, on its merits alone, will stand the test of time. It is the only known concept capable of furnishing provable solutions to all planetary anomalies.

2. The paper was already too long to leave room to discuss the many implied deficiencies of current beliefs; e.g., Laplace's accretion concept and any interpretation derived via Poe's 1848 Big Bang. Both concepts are widely known and needed no comments other than those implied by the LB/FLINE model. An honest effort was made to stay within the bounds of the Four Laws of Planetary Motion proving the fiery dynamic origins of solar systems and, subsequently, the five-stage evolution of planets (including exoplanets) and moons.

3. Quite to the contrary, my ideas were not assertions of fact without logical proof or any supportive observations. Logical proof resides in the Seven Diagrams clearly revealing how each nebulous planetary mass attained its original and current orbital spacing — no other concept can offer a definitive solution to this mystery. These Diagrams required 15 years and much logic based on a research background and the ability to separate myth from fact. Once in orbit, fiery spheres follow the second law of thermodynamics as they cool through their five stages of evolution. Mankind has only to look to the skies to observe each stage of planetary evolution. Convincing proof of the Four Laws and the subsequent planetary evolution resides in the absolute fact that internal nucleosynthesis is essential to evolution; one cannot exist without the other. The supportive observations of planetary evolution via internal and external discoveries of all planets are voluminous. Everything evolves continually and in full accord with natural laws; there are no known exceptions. Logic and voluminous supportive evidence in my book, *The Spacing of Planets: The*

Solution to a 400-Year Mystery, provides sound assurance that the FLINE model contains no fundamental flaw.

The proven origins, evolution and close relationship of hydrocarbon fuels (1973-1980) is a facet of the FLINE model — an example of a precise and logical concept with voluminous supportive evidence. It is unfortunate that this definitive LB/FLINE model, again as with Copernicus, offers such powerful evidence for a viable alternative to prevailing beliefs. In any era of science, the greatest joy is discovery of a revolutionary truth; but tragedy too often resides in the reasons for, and the consequences of, its suppression. If anyone in the RAS can find a fundamental flaw in this new FLINE concept, please make me aware of it.

Belief vs. Science: Which is More Important? (April 2001)

On the evening of March 24, 1999, Stephen Hawking, the author of the highly successful book, *A Brief History of Time*, was in Atlanta to receive the American Physical Society's Lilienthal Award and to give an address titled *The Universe in a Nutshell*. According to the follow-up news article, "the ideas discussed were no longer new and are not universally accepted. But his talk was less an outline of new ideas than a public event featuring the famous scientist."

According to the report in the Atlanta Journal-Constitution, "many physicists question whether Hawking's highly mathematical, highly speculative work will produce the unified field theory he and others have sought, combining Einstein's General Theory of Relativity (which deals with the largest physical phenomena in the universe) and quantum mechanics (the behavior of the most elementary particles). But they all agree that no one has looked further into the nature of space and time. ...*A Brief History of Time*, his first book for lay readers, made him a celebrity beyond the confines of theoretical physics."

The news report continued: "The question, among some physicists, is whether cosmology is actually physics, whether it is science or science fiction. Its hypotheses are conjectual when pressed to the limit, and its conclusions can be neither proved nor disproved. Hawking, himself, has acknowledged the dilemma. 'Cosmology isn't science,' he told *Discover* magazine, 'if it doesn't predict what we can observe.'" [Viewed in the LB/FLINE model's non-conjectual version, cosmology does qualify as science].

To pursue the issue further, a quote from *The World and I* magazine (Mar 2001, p. 251) seems in order: "A point that keeps appearing throughout Ingram's essays is the subjectivity of science. We scientists sometimes fool ourselves into thinking that our work is always objective. It is instilled in us that our experiments must be totally unbiased. They must be double-blind, randomized, and controlled. 'There is an object lesson here for anyone out there who still believes that science is an unbiased route to the truth,' says Ingram. It's not that scientists are dishonest; it's just that they start with different preconceptions. 'After all,' he says, 'the answers you get are constrained by the questions you ask.' Different scientific approaches can lead to different versions of the truth. Sometimes it's belief, not science, that is most important, 'until the data are overwhelmingly persuasive.'" One of the best examples in which the data are overwhelmingly persuasive is the issue of the true origins and evolution of our hydrocarbon fuels: gas, oil, and coal via internal nucleosynthesis and polymerization; its many facts are irrefutable.

Ingram hits on another basic truth about science and society: that it is difficult to interest people in learning about science unless you give them a practical reason for knowing it. At this writing (April, 2001), the new energy crisis now threatening our economy stems directly from the antiquated fossil fuels hypotheses of the 1830s and 1920s: that gas, petroleum and coal are products of decay, marine sediments and plants, respectively, and thus abound in very limited quantities. Nothing could be further from the truth. The FLINE model's many provable facts about the origins, evolution and close relationship of these fuels should, and would, excite the public if permitted equal treatment by the news media and scientists, with financial backing equal to that of the antiquated fossil fuels hypotheses. Such fair treatment would go a long way in keeping any potential energy crisis at bay — and would give people a practical reason for learning about science. In any era of science, the greatest joy is discovery of a revolutionary truth; but tragedy too often resides in the reasons for, and the consequences of, its suppression.

Understanding the origins and evolution of hydrocarbon fuels can open many doors to understanding a cosmology science that is capable of predicting what we can observe. But first, one must be willing to let go of all antiquated beliefs and preconceptions about origins, while investigating the newer concepts. The results of all discoveries must be interpreted in perspectives newer and more factual than Poe's antiquated Big Bang. At this time, belief still reigns over science — but who knows for how much longer?

Letters and Memos

To: Glenn Strait, The World and I, Washington, DC. Jan 8, 1998

Thanks for the press release dated 98-01-05 concerning the Lunar Prospector mapping the magnetic patchwork of the Moon, and for the statement that you are beginning to understand the FL/IN theory. The revelations in the press release, as well as other space findings (on Mars, etc.), continue to corroborate this new paradigm of planetary origins.

The patchwork of weak magnetic fields on the Moon's surface are remnants of the strong magnetic field inherent in the energy mass at the birth of the Moon. As you may realize, all planets and moons began as fiery masses of nuclear energy with powerful electromagnetism inherent in each mass. This has been confirmed by the proposed Fourth Law of Planetary Motion, the principles of planetary evolution via natural laws (including the Laws of Thermodynamics), the recent findings on Mars (similar surface magnetic fields, rocks, dry river beds, volcanoes, etc.), and other space discoveries.

This Moon project will simply add more confirmatory evidence of the prior existence of the sphere's original strong magnetic field that grew ever weaker as its core energy transformed into atomic matter that initially encapsulated it via atmospheric gases (e.g., see large planets), followed by crustal formations. This is the evolutionary pattern of all planets (including exoplanets, moons, comets and asteroids (a special case). All the smaller planets of our Solar System should now show similar patterns of remnant magnetism. And perhaps most of the irregularities of the remnant magnetism can be attributed to the last effects imposed on the surface by the dying energy core — via its final-stage gasps, so to speak.

The discovery that "electrons were coming from the [Moon's] surface" raises some questions about their source. Do they come from outer space or from inside the sphere? My guess is the latter source rather than the solar wind source suggested in the article. Here we have two tentative conclusions, each based on a different perspective of the Moon's origin and evolution. Can we look at Earth's field of electrons (in both perspectives) and find the correct answer? Perhaps they stem from both sources.

I agree with the statement that: "Magnetic field information also could provide constraints on the physical processes undergone by the Moon in its evolution, such as how the core formed, the thermal evolution of the crust, tectonic processes and erosion." This line of reasoning, if followed, inevitably will bring scientists ever closer to the FL/IN paradigm. Once this goal is reached early in the 21st century, the answers to the remaining anomalies of solar systems will fall into place, and science will benefit immensely from this breakthrough. Two crucial keys will be the discovery and recognition of the true nature of comets and asteroids. I'm convinced that the FL is the most monumentally significant discovery about our Solar System since Kepler's discovery of the First Three Laws of Planetary Motion.

Glenn, at what point along the way would it be feasible to publish the FL/IN paradigm as a serial article in your magazine? It does seem to me that with a scoop of this magnitude, at this stage of development, and with the persistent encroachment of confirmatory discoveries, time is of the essence.

P.S. A paper on the solution to the Fourth Law is being submitted next week to a leading journal.

To: Glenn Strait, The *World & I* Magazine, Washington, DC 20002. 01-26-98

One night last week, I watched with trepidation as a science program about comets unfolded. The plan is to land a small drilling mechanism on a comet and bring back samples for analysis. In the film, scientists already were designing the sampling mechanism. I'm not sure of the target date for the landing — it could be around 2007 or later.

The plans are based on the assumption that comets are 'dirty snowballs' — frozen leftovers from the creation of our Solar System. However, the evidence obtained during my research of the past 25 years on the subject of planetary origins clearly indicates that comets are fiery masses of nuclear energy. If true, the consequences of landing on such a mass will be monumentally significant, to say the least. If triggered, the explosion would be similar in force to those of the fragments of Comet SL9 on Jupiter in July, 1994. But better a mechanism than a person.

However, in June, 2000, the Deep Space 1 probe is scheduled to fly past Comet West-Kohouteklkemura, taking close-up pictures and studying the coma surrounding the comet's dark nucleus. It may also view the two Martian moons, Phobos and Deimos, which I believe are actually shells of burned-out comets.

In view of this information, it seems wiser to sample one of these moons rather than the comet. Certainly, it would be safer and easier to obtain samples representative of matter produced by a comet to form its dark shell. But, of course, this would not settle the issue of the true nature of comets; it is perhaps mandatory that the landing take place as planned to verify their nature. I only wish it could be done sooner.

A more elaborate explanation of comets can be found in my book *THE SPACING OF PLANETS*. Herewith are copies of pages 99-108.

To: Editor Walter Anderson, *PARADE* Magazine, New York, NY. May 11, 1998

I would like to comment on David Levy's article, *Was Chicken Little Right?* concerning comets and asteroids.

Substantial evidence reveals that a comet consists of a small shell containing an extremely powerful mass of energy capable of exploding instantaneously upon contact with anything as thin as the outer atmosphere of a sphere. Examples are (1) the 21 explosions of fragments of Comet SL9 at Jupiter's cloud-tops in July, 1994, and (2) the upper-atmosphere explosion of the 1908 Siberian comet. Asteroids, rather than comets, create the type of huge 100-mile diameter craters described by Levy. But such craters are far more often created by ejecta forces from within a sphere — which accounts for the multitude of craters on Mercury, Callisto, our Moon and on surfaces of other spheres.

Cogent evidence of the true nature of comets can be found in *COMETS* by Donald K. Yoemans (1991), and in my books: *Undermining the Energy Crisis* (1979), *New Concepts of Origins: With White Fire Laden* (1986), *The I-T-E-M Connection* (1991), *The Spacing of Planets: The Solution to a 400-Year Mystery*. My books also differentiate between comets and asteroids: comets are fireballs of "Sun-like" matter (Descartes' description of planetary cores) - actually Sun-like nuclear energy/matter.

Cogent evidence clearly reveals that Asteroids are fragments of a disintegrated (actually 3) planet(s), just as Olbers stated in 1802. They can, and do, create large craters by crashing into larger masses. However, these are very rare events and, contrary to popular dogma, such collisions are not the cause of the vast numbers of craters on Mars, our Moon, Callisto, etc. As proven by Allan Mills in the 1980s, the vast majority of such craters were caused by ejecta gases (and other matter and energy) from within each planet and moon. In turn, the causes of the ejecta forces are detailed in my aforementioned books.

Since each planet is a self-sustaining entity that creates its own atmosphere, crust, water, life, etc., via internal nucleosynthesis (IN), there is no need to look to comets or other outer space objects for sources of such matter. The key factor in the evolution of planets is that internal nucleosynthesis (IN) and evolution (E) are inseparable: one cannot exist without the other. This is the basic underpinning of the proposed FL/IN/E paradigm of origins that is now strongly supported by my proposed Fourth Law of Planetary Motion (FL), detailing how the fiery masses that evolved into planets attained their geometrically-spaced orbits around the Sun and why the last two planetary orbits are unorthodox.

The comet crashes on Jupiter were indeed "the most incredible explosions ever seen on another world." Each powerful explosion far exceeded all prior supercomputer calculations that were based on the "snow-ball" hypothesis. Instead of the predicted small, orange flashes one fiftieth as bright as Jupiter, the huge, white-hot flashes were up to fifty times brighter than Jupiter. Fragment G alone yielded at least 6 million megatons of energy. To expend an equal amount of energy, a Hiroshima-type bomb would have to be detonated every second for 10 years! These powerful explosions cast serious doubt on the "snowball" comets hypothesis while conclusively confirming my 1970s prediction of their true composition: nuclear energy. Supportive evidence of the latter removes the remnants of any doubt.

The several iridium layers (and all other layered matter throughout geological history), including the layer marking the demise of the dinosaurs, can be explained far more logically by the FLINE paradigm than by the alleged collisions of missiles from outer space. Substantive evidence proves that the dinosaurs actually died out over a five-hundred-million year time frame as a result of atmospheric changes attributed to a dynamic, evolutionary Earth driven by the energy of its internal engine.

While the prevailing hypotheses make exciting reading and movies, they do grave disservice to true science, to present-day science students and to posterity. Ironically, the more factual truth about our planetary origins and evolution creates even more exciting scenarios.

True science should remain open to presentations of both sides of any controversy. It is indeed unfortunate that a naive public hears and sees only the less factual side of these issues rather than the full story.

To: Glenn Strait, The World & I, Washington, DC 20002   07-03-98

Thanks for the news release from Seattle's University of Washington's Professor Donald Brownlee, forwarded to you by Steve Maran, American Astronomical Society (AAS). I noted with interest that Maran disclaimed endorsement by the AAS; unfortunately, the reason was not specified.

I am disappointed that the Stardust craft will pass only 75 miles from the main body of the comet, Wild 2, rather than attempting to land on it. It is not likely that capturing samples of the dust particles and ionized gas from the comet will reveal anything beyond what is already known about the composition of cometary ejecta. The problem is, and always has been, in interpreting the data obtained from previous analyses of such ejected matter. When correctly interpreted, the findings argue strongly against the 'dirty snowball' belief, while offering even stronger evidence favoring the fireball concept.

The most effective way to gain any real insight into the true nature of the comet is to carefully measure the temperatures in and around the nucleus and the coma. However, the mission is being run on the false assumptions that "comets are about equal parts ice and dust", and thus, "the particles will be cryogenically preserved interstellar dust left from the birth of the solar system. ...such grains can be found only in the outer solar system because heat has destroyed them nearer the Sun." How does Brownlee compromise the contradicting fact that many comets have the Sun as one focus of their orbits?

Dust particles gathered from outer space and from different levels in our atmosphere naturally coalesce into larger particles, which is the nature of such lightweight matter. However, this reasoning cannot be carried over to apply to massive boulders (asteroids, etc.) and comets coalescing to form larger masses. In real life, such larger spheres are formed in full accord with the FLINE paradigm's three chronological, inseparable and ongoing events: The Four Laws of Planetary Motion (FL), internal nucleosynthesis (IN) and evolution (E). In any active sphere, IN and E are inseparable; one cannot exist without the other. In our Solar System, the Four Laws of Planetary Motion must be added as the initial (and ongoing) factor that accounts for all motions of the planets, including their orbital spacing, thereby adding substantive evidence to this revolutionary paradigm (see enclosed flyer for more details).

At least Brownlee got the name right: Stardust — truly the ejecta spewed from every fiery star, sun, early-stage planet and active comet throughout the Universe can be called stardust. However, its analyses are misinterpreted continually as force-fitted evidence that comets are pristine snowballs rather than the fireballs in which the virgin dust particles and ionized gases are created via internal nucleosynthesis. As revealed by the extremely powerful explosions of Comet SL9 at Jupiter's very thin cloud-tops in 1994, newly-formed comets are easily triggered into nuclear explosions. Rather than "the preserved building blocks of the outer planets", comets are ejecta energy from larger masses of nuclear energy comprising all stars. Perhaps Brownlee will "find clues about the formation of the Solar System and perhaps of the Universe itself", but they must be interpreted accurately. Monitoring the temperatures and, perhaps, triggering an explosion of the comet could add immeasurably to these interpretations.

In time, why the AAS withheld endorsement of the project will become clearer. The results of the $200 million mission could alter the direction of scientific thought on the origins and evolution of planets.

To: Glenn Strait, *The World & I* Magazine, Washington, DC 20002. 07-08-98

Thanks for the information received today on our Solar System's hottest planetary surface: Io — truly exciting news. In answer to your questions concerning the fiery nature of Io versus the quiescent nature of our Moon (and Mercury) (three spheres of approximately equal size), the FLINE paradigm offers this reasoning:

Planets evolve through five common stages of evolution in accordance with size. However, our Moon and Mercury are entering the fifth and final stage (inactive), while Io is only in the fourth stage (rocky), but is not as far along in this stage as Earth's evolution. Io's appearance, activity and high temperature now are similar to those of Earth during that specific time frame for the early part of our planet's fourth stage of evolution. With this information in mind, logic dictates that some factor(s) in addition to size played a crucial role that accounts for these differences in the stages of evolution of these three spheres.

The options: (1) Jupiter is playing an active role in creating tidal friction heat in Io; or (2) Jupiter played some other crucial role in the past. The first option represents the prevailing belief, *which can be readily nullified by a simple analogous comparison of the Jupiter-Io system to the Sun-Mercury system in which no tidal friction heat is created in Mercury*. The second option fits neatly into the FLINE paradigm: Io is much younger than both our Moon and Mercury; in reality, it is the youngest moon of Jupiter and is in the smallest orbit of Jupiter's four large moons. As such, we can reason that a later birth of Io accounts for its later start in the evolutionary stages, and, thus, its current lag behind similar-size spheres of our Solar System in evolutionary time frames. But how and why was Io born at a later time than our Moon (and Mercury)? And is there evidence to support this theory?

Based on substantive evidence, the FLINE paradigm teaches that Earth and Moon are contemporary spheres, while Jupiter and Io are akin to a mother-and-child relationship. Some time after the original layout of the Solar System (in full accord with the Four Laws of Planetary Motion), and after Jupiter had entered its second (gaseous) stage of evolution, the original energy mass we now call Io was ejected from Jupiter's gigantic energy core and sent into orbit around, and close to, Jupiter. After this concept was visualized, my reasoning was that a relationship could exist between this ejecta and Jupiter's Great Red Spot. *Does the Spot mark the exit of the Io mass? And is the stormy Spot still driven by the same forces that caused the ejection and orbiting of the Io mass? If so, the orbit of Io should pass over or nearly over the Spot. A picture discovered later in one of my library books clearly shows this to be the case.* Herewith is a copy of the Voyager photo showing this very close relationship. Someday Io's later birth will be confirmed by testing its oldest rocks.

Glenn, there are large numbers of scientists who could benefit from the knowledge and opportunities offered by the FLINE paradigm revealing the relationship of its three inherent and inseparable factors: The Four Laws of Planetary Motion, internal nucleosynthesis and evolution. In its perspective, all anomalies appear solvable. If you should decide to publish any facet(s) of the paradigm, I would be happy to work with you. Meanwhile, I ask that you keep this information confidential; it is yet to be added in the next update of my book (ninth edition) on planetary origins and evolution.

To: Glenn Strait, *The World & I* Magazine, Washington, DC.   06-23-99

Thanks for the information on comets as described by the AAS press release of 17 June 1999. I'm always amazed at the (mis)interpretations of advocates of the version of comets as dirty snowballs, and the tenuous manner in which the results are presented. In this article, I counted 11 examples: "may want", "indicate", "was thought", "could have formed", "indicates", "researchers believe", "may be", "probably formed", "believed to be", "are probably", "may reveal". Speculations are necessarily abundant, and they do breed fast.

A relevant question that comes to mind concerns the 21 violent explosions of Comet SL9 at the thin cloud-tops of Jupiter in 1994. The best supercomputers and other computers had predicted that the snowball impacts would be relatively small. Surprisingly, the tremendous explosions far exceeded those expectations. Since the actual results had been accurately predicted (by written submission) by the FLINE model as the greatest explosions ever witnessed in our Solar System, I was on "cloud nine". Could snowballs, or anything other than nuclear energy fireballs have caused such massive explosions? NO!

Backed by the Four Laws of Planetary Motion, the FLINE model clearly reveals that all planets began as fiery energy masses. In the early seventeenth century, Willebrord Snel recognized that comets formed from fiery exhalations from the Sun, and glowed by their own light. Observations reveal that comets are short-lived, in the range of a few hundred to a few thousand years before burnout, explosions or impacts onto larger bodies, usually the Sun, from whence they came.

To explain the data, per your request: You are familiar with the IN-E relationship of the new model in which all atomic elements are created via internal nucleosynthesis (IN) that is absolutely essential to the processes of evolution (E). Just as evolution is not possible without an energy engine to drive it, the creation and exhalations of the massive amounts of CO from Comet Hale-Bopp would not be possible without its internal processes of nucleosynthesis. (CO, a common product of fiery sources, is only one among many compounds produced by this fireball and other comets). All comets vary in the type and amounts of their virgin matter because of combinations of variables: size, temperature, pressure, distance, etc. These exhalations can be readily observed emanating from the multi-port, jet-black shell of the nucleus, even when comets are far beyond the effects of the Sun's heat. Of course, production rates and types of the multi-jet streams are affected by distance from the Sun. One amazing fact is that *Comet Hale-Bopp* <u>blazed away</u> <u>beyond</u> <u>the</u> <u>orbit</u> <u>of</u> <u>Jupiter</u> <u>25,000</u> <u>times</u> <u>brighter</u> <u>than</u> <u>Comet Halley</u> *did when it was at the same distance from the Sun* — a spectacle reminiscent of the early appearances of Comet Halley between 12 B.C. and 1682 when its former brilliance caused a sensation in the sky every 76 years.

The finer points are covered in my book, *The Spacing of Planets: The Solution to a 400-Year Mystery*. I believe you have a copy; if so, refer to pages 96-108 and 126 for complete details, examples, photos, etc., of comets. Like Descartes, Snel was right; unfortunately, neither knew about the internal nucleosynthesis in the fiery masses: stars, suns, planets, brown dwarfs, etc., throughout the Universe, and that all such masses came via explosions of larger masses of energy. As you know, nucleosynthesis is the transformation of energy into matter ($E=mc^2$) under conditions of high temperatures and pressures within a nuclear energy mass. Such atomic matter can be observed on all fiery masses, including our Sun, stars, planets, moons, comets, etc. In all these cases, active E is possible *only* via active IN.

I hope this answers your request of how I explain the data presented in the news release. In summary, the FLINE model's Four Laws of Planetary Motion (FL) placed the fiery planetary masses in GR-spaced orbits during their initial stage in which internal nucleosynthesis (IN) begins driving their five-cycle evolution (E). Comets are smaller versions of these processes; CO is an IN product common to both.

To: Glenn Strait, *The World & I* Magazine, Washington, DC 20002   07-07-99

Thanks for the AAS News Release received yesterday concerning Comets Wirtanen, Hale-Bopp and Halley. The two common threads throughout are: (1) the relationship between the comet's distance from the Sun versus coma size; and (2) the jet-black color of the nucleus.

In the perspective of the snowball belief, the standard explanation of the increase in coma size as the comet approaches the Sun is well known. But several questions persist: How can a coma consisting of hydrogen up to one million miles in diameter form a spherical envelope that continues to grow around such a tiny nucleus when approaching against the Sun's solar wind? Why does the coma remain so large at such great distances from the Sun — well beyond its warming effect? For example, Comet Halley's coma was almost 500,000 km at a distance of 8.5 AU from the Sun (almost to Saturn), and Comet Hale-Bopp's coma at 8.66 AU was more than a million km across, or nearly ten times larger than Saturn!

Why do ejecta forming the hydrogen coma continue at these great distances? If a snowball, why is the nucleus always jet-black? Will Comet Wirtanen's nucleus at its aphelion show through the telescopes as a black dot or a brilliant dot?

When comets are viewed as energy masses, or fireballs, as depicted in the FLINE model of planetary origins, logical answers to these questions simply fall into place. Too long for this letter, they can be found in my book, *THE SPACING OF PLANETS: The Solution to a 400-Year Mystery* (Chapter IV: *Moons, Planetary Rings and Comets*, beginning on page 87).

One last comment: I was relieved to learn that the missions to comets do not include a plan to try to land a man on Comet Wirtanen. From the viewpoint of safety and discovery, the instrument landing, per se, should be exciting enough.

Lunar Prospector Findings on the Moon's Metal Core.

In the AAS press release of August 10, 1999, the authors are in agreement with the FLINE model in their conclusion that the Moon's core is smaller than the cores of the inner planets of the Solar System (SS). One principle of the new model is that the sizes of all cores of planets and moons created simultaneously as a solar system are proportional to their total mass: the smaller the mass, the smaller the core. At the other end of the scale, the largest planet, Jupiter, at our SS's geometric mean, has the largest core. The initial reasons for the various sizes of the planetary masses (and their corresponding core sizes) are explained by the forces acting on the decelerating smaller mass as it acted in unison with the Sun during the dynamic layout of our Solar System in full accord with the Four Laws of Planetary Motion some five billion years ago. Size and distance from the Sun are the two principal factors explaining the compositional differences among the planets and moons as they evolve(d) through their five common stages of evolution. Meanwhile, the core of each planet dwindles as a function of time — a consequence of the slow transformation of its energy mass into the sphere's compositional matter. The core-size evidence in the press release strongly supports the FLINE model.

Unfortunately, the core size and composition of the Moon are the only conclusions in agreement with the FLINE model. The conclusion drawn from the results of the magnetic tests is premature: how can one differentiate between the actions of the magnetism of a small metal core and a small nuclear energy (or plasma) core solely on the basis of these findings? In the FLINE model perspective, whether this evidence is indicative of remnants of a small nuclear core or of a metal core remains debatable. If researchers would only consider possible scenarios for interpreting their findings!

On previous occasions, we have discussed the highly speculative hypothesis that a Mars-sized object, collided with Earth and bounced away to become the Moon. Were all the moons of other planets made in similar manner? If so, how could so many moons have bounced off the thin, deep clouds of Jupiter? The FLINE model teaches that all spherical moons were created and evolve(d) in a manner like that of the planets. As one scientist put it, *"Books and articles supporting this giant impact hypothesis of lunar origin are more a testament of our ignorance than a statement of our knowledge."* Full details can be found on pages 87-94 in my book, *The Spacing of Planets: The Solution to a 400-Year Mystery*; they present powerful arguments for a different and definitive origin of our Moon and other moons.

To: Glenn Strait, *The World & I* Magazine, Washington, DC 20002.   11-22-99

Thanks for the *AAS Press Release (Nov 17); Jupiter May Have Moved in Toward Sun, Probe Data Suggest*. My thoughts are that two responses would be helpful: one to you as well as the one to the researchers, as suggested, through you, and enclosed herewith. Please feel free to improve its composition where feasible.

It was encouraging to read that "the Galileo spacecraft's suicide probe as it plunged into Jupiter's roiling atmosphere in December, 1995 has stamped a huge question mark over the prevailing models of how our solar system began." In my 1995-1996 book, *The Spacing of Planets* (p 67), I had written a similar message: "Just as Galileo himself drove the dagger into the heart of the Ptolemaic version by promoting the Copernican idea of a Sun-centered Solar System, the Galileo orbiter is plunging a dagger into the heart of the prevailing dogma about planetary origins." It is gratifying to learn that scientists are now beginning to question prevailing models —the first critical step toward understanding and eventually recognizing the truths inherent in the FLINE model of planetary origins and evolution. The next step will be the realization that the Accretion hypothesis is fundamentally flawed and should be discarded from textbooks and replaced by the more logical and factual model.

Your stated realization that the reported data fit in well with the FLINE model is good news. Indeed, the large quantities of heavy noble gases in Jupiter's atmosphere are precisely as predicted and explained by this revolutionary model for any planet in the second stage of planetary evolution via internal nucleosynthesis (IN). While the overwhelming evidence for IN is recorded throughout the book, it is concentrated more in Chapters II & III — some 42 pages of what appears to be irrefutable evidence of IN.

The Fourth Law of Planetary Motion, explained in Chapter I of the book, shows that Jupiter has moved only 1.46 AU, or 21.3%, closer to the Sun since its origin some five billion years ago — not enough to be a factor in the speculations of the researchers concerning its orbital changes or movements from far distant places. As based on the Accretion hypothesis, the illogical explanations offered in the report explaining Jupiter's origin and orbital changes call for more and more speculations. It reminds me of the tale about the three blind men, each feeling and describing a different part of an elephant, and each arriving at a much different — and entirely wrong —conclusion about it.

One must ask how long this situation can go on before scientists do seriously "question...prevailing models of how our solar system began." None of the wild speculation in the report would have been necessary if only they had been advocates of the viable alternative: the definitive FLINE model of planetary origins whereby all of Jupiter's anomalies can be explained easily and proven conclusively without the need for speculation. However, it does appear that more and more scientists are leaning in its direction, whether or not they have heard about my writings of the past 26 years.

After 22 months, my manuscript on the solution to the proposed Fourth Law of Planetary Motion is still in the hands of referees at the Royal Astronomical Society, London. I believe that if they could find something fundamentally wrong with it, an early (and eager) response would have been received. One can only wait and hope.

To: Brig. Gen. Gordon C. Carson III, Livingston, TX.   Feb 3, 2001

Thank you for your letter of January 28 concerning the readings about the old and new concepts of origins and evolution, and for your words of encouragement to keep up the good work.

The responses you have received from colleagues reading *The Spacing of Planets* sound much as expected: "mostly good and some not-so-good." I imagine the good comments came from readers who read with open minds void of any strong bias favoring prevailing beliefs. The not-so-good comments probably came from strong believers in the popular Big Bang/Accretion version of universal origins. This, of course, is understandable when we look at the history of significant breakthroughs in opposition to beliefs of their time. But time and facts remain steadfast and powerful allies of truth.

I would be genuinely surprised if any of the unfavorable comments contained a valid rebuttal of any facet of the LB/FLINE model. If so, it would be the first one since conception of the initial facet of the concept in 1973: a theory solidly constructed with irrefutable evidence against the erroneous concept of 'fossil fuels' and in strong favor of the energy fuels concept of the origins and close relationship of these hydrocarbon fuels. In the past, the only rebuttals given were simply, 'I don't believe it' — always said before showing a willingness to take the time to learn about its volume of supportive, intertwined facts. The concept needs a rebuttal or two if any weak points in the LB/FLINE are to be found and strengthened or clarified. If you do receive a definitive one, please forward it to me for this essential purpose.

Once one understands the basic principles of how these fuels came into being, it is relatively simple to understand how all matter came into being via the same principles. Simply by applying Nature's laws of physics and chemistry and correctly interpreting observations and relevant data, one can arrive at pieces of the puzzle that always interlock precisely with the previously fitted pieces.

Unfortunately, proponents of the Big Bang/Accretion beliefs remain firmly in control of the mind-set of most scientists and the general public. That hold isn't likely to diminish in my time. To paraphrase the astute observation of a great physicist, Max Planck, science must advance slowly, one funeral at a time.

Not having a Ph.D. with a mind trained in prevailing beliefs did offer a big advantage in being able to see things in new perspectives. Chemistry and chemical engineering degrees, along with my career in industrial R&D (applied), problem solving, etc., boosted and broadened that tremendous advantage, without which I would not have been capable of fitting all the pieces together — even with a Ph.D. Fate did play a major role.

To: Glenn Strait, The World & I Magazine, Washington, DC 20002.   03-12-01

Thanks for the four e-mails concerning Comet Hale-Bopp, Ganymede, Io, and buckyballs. The printing from this end was faulty, but I will try to decipher it well enough to supply answers in accord with the FLINE model. Meanwhile, Kay is installing a new printer that better matches her new computer.

The news release that Comet Hale-Bopp, now at a distance halfway between Saturn and Uranus, retains its huge coma and is still spewing trails of dust, gas and chemicals — precisely as predicted and explained by the FLINE model — is truly exciting (Ref. *The Spacing of Planets*, p. 103, (1996). Quoting the primary reason: "Without the continuous activity of nucleosynthesis in these distant comets, the existence of comae and the production of gases and dust would not be possible [at great distances from the Sun]. Perhaps some day soon, improved detection methods will enable astronomers to detect and verify that...gases and matter do continue spewing jet-like from comets in the deep cold dimness of the outer SS — just as predicted in 1973 by the IN theory. This will be another giant step toward proving the FLPM/IN concept of our origins." Comets are fireballs, not dirty snowballs; they played no vital role in the formation of planets.

The interpretations researchers give for the observations of Ganymede's surface features are bringing them ever closer to the FLINE model. But they need to go a giant step further by identifying the driving force behind these ongoing changes in its surface features: the fiery energy core, the nucleosynthesis engine that basically drives all planetary changes, volcanism, evolution, life, etc. The same principles quite naturally apply to younger Io, whose dramatic differences are explained by the new model.

Becker *et al.* should consider how buckyballs are made in the carbon-black industries that furnish carbon-black to tire manufacturers and other industries — and we live in a carbon-dominated world. Earth's internal manufacturing system is quite capable of making buckyballs containing trapped noble gases; there is no need to claim that they came from outer space. To do so, researchers are making the same mistake made by proponents of the belief that our iridium layers came from outer space, or that the Moon's titanium came from outer space. Quoting again from *The Spacing of Planets*, p. 93: The titanium produced on the Moon is analogous to the tellurium produced on Venus, the iridium produced on Earth, the magnesium produced on Neptune, and the abundant metals observed at the impact site of the G fragment of Comet SL9 on Jupiter — all are in situ ejecta products of nucleosynthesis." If from outer space, these metals would not have been so selective of landing sites; each is the result of internal conditions that prevailed during each sphere's time-frame for making and ejecting a specific chemical.

In addition to the manuscript on the solution to the Fourth Law of Planetary Motion and its FLINE model, which was mailed to you on March 12, I have a short manuscript on *Comets and Asteroids: Keys to Our Origins?* that details the complete story about their true nature, origins, etc. If you would like a copy for consideration to publish, just let me know.

A Question of Ethics in Science

The third part of a series of four videos on *Savage Planet*, entitled *Deadly Skies* was shown on GPTV, PBS on November 19, 2000, at 7:00-8:00 p.m. The featured astronomers included David Levy and Carolyn Shoemaker as experts on asteroids and comets.

For the first time ever, the presentation actually attributed the source of asteroids to an exploded planet that once existed between Mars and Jupiter. Since that event — the most powerful explosion in the history of mankind — the resulting fragments have been falling periodically onto Planet Earth for an unknown number of years.

This sudden about-face conclusion of scientists raises questions: Are the asteroids a product of the Big Bang (BB) (as has been the standard claimed for some five decades) or are they actually remnants of a disintegrated planet, as first stated by Olbers in 1802 — and a fact that has been a vital principle of the evolving LB/FLINE model since 1980? What do the asteroids tell us about the origin of our Solar System? Do planetesimals (asteroids) really accrete into planets or do they break into smaller pieces over time (as stated in the video)? Why this sudden acceptance of asteroids as remnants of an exploded planet after so many decades of teaching and promoting their creation via the gaseous-dust accretion concept? Could such original gas and dust have evolved previously from the alleged Big Bang?

The answers to these questions, as supplied by the LB/FLINE model of origins during the past two decades, speak loudly and clearly against Poe's BB hypothesis (1848) and its inherent, but antiquated, Accretion concept initiated by Laplace in 1796. Simultaneously, the model proves that Olbers was correct in his assertion that asteroids are fragments of an exploded planet (actually three spheres). Remember that the fragments are clearly in three definitive orbits between Mars and Jupiter. The fact that scientists now accept another facet of the LB/FLINE model (the true nature and source of asteroids) is a giant step in the right direction. The next giant step will come when they eventually recognize and admit the true nature and sources of fiery comets.

The second surprise came when the video showed much more powerful explosions of Comet SL9 on Jupiter than had ever been depicted in previous programs on the subject. To their credit, they did not yet go so far as to admit that these powerful explosions were of a nuclear nature (as proven by the FLINE model), but the enhanced pictures gave that impression. Eventually, astronomers will be forced to admit the nuclear nature of comets, and to understand the sources of their ongoing origins. The acknowledgement of their nuclear nature, along with this long-overdue recognition of the true nature of asteroids, clearly will undermine both the BB and the Accretion hypotheses.

Of all the questions raised here, one issue stands out conspicuously: Why the sudden switch from the belief that asteroids are early accretion products of the BB to the belief that they are remnants of an exploded planet? Why does science continue to ignore and reject the LB/FLINE model in spite of its substantiated evidence and its impeccable record of accurate predictions during the past two decades?

Deadly Skies and similar TV programs continually bring scientists ever closer to the LB/FLINE model of origins and evolution. The question of ethics in science eventually must arise: At what point will it become obviously unethical to continue ignoring the LB/FLINE model described in this book and in previous publications of the past two decades, while continuing to "discover" its inseparable principles of creation via ongoing universal nucleosynthesis and its accurate interpretations of relevant data; e.g., the nuclear explosions of Comet SL9 on Jupiter, predictable neither by any other known concept nor by the world's most powerful supercomputers? Is this encroachment method ethical or unethical science?

Addendum

The greatest obstacle to advancement in the sciences of origins and evolution is the illusion of knowledge.

If you had to identify, in one word, the reason scientists have not achieved their goal of understanding universal origins and evolution, that word would be "consensus".

In science, the greatest joy is discovery of a revolutionary truth; the saddest tragedy resides within the reasons for, and the consequences of, its suppression.

Alexander Scarborough, Jan-Mar, 2001

Alexander A. Scarborough

CHAPTER V

A BIG BANG OR LITTLE BANGS?

"When we try to pick out anything by itself, we find it is hitched to everything else in the universe."

John Muir

The Three Key Observations

Three observations provide the fundamental basis for the standard cosmology featuring the Big Bang (BB) theory:

1. The observed expansion of the Universe (usually interpreted in the framework of relativity as an expansion of the metric of space).
2. The 2.726 K cosmic background radiation (CBR), interpreted as a remnant of the BB.
3. The apparently successful explanation of the relative abundance of the light elements.

In reality, the same observations serve equally well as the fundamental basis for a different concept: the Little Bangs (LB) theory that intermeshes precisely with the geometric origin of the SS and the evolution of planets via nucleosynthesis. Pieced together during the past 28 years (1973-2001), this revolutionary perspective surprisingly offers answers to the many questions that pose challenges to the BB theory. The two concepts need to be examined closely with open minds.

Questions About an Expanding Universe (1992)

Hetherington's *Encyclopedia of Cosmology* summarizes the status of the BB theory: "...problems have been numerous and not all of them are solved to the satisfaction of all cosmologists. In particular, the large-scale homogeneities observed in the 1980s seem to indicate a structured Universe, which may contradict one of the foundations of the BB cosmology, the uniformity postulate (or cosmological principle). This and other problems have recently caused some cosmologists to declare the BB theory in a state of crisis. However, since no plausible alternative exists, the almost universal belief in the BB model has not been seriously shattered."

The reason for the fallacy of the last statement is best summarized in the article, *Why Only One Big Bang?* (*Scientific American* Feb 1992), by Geoffrey Burbidge, a professor of physics at the University of California, San Diego. To quote, "Those of us who have been around long enough know that peer review and the refereeing of papers have become a form of censorship. It is extraordinarily difficult to get financial support or viewing time on a telescope unless one writes a proposal that follows the party line. A few years back, Halton C. Arp was denied telescope time at Mount Wilson and Palomar observatories because his observing program had found, and continued to find, evidence contrary to standard cosmology. Unorthodox papers often are denied publication for years, or are blocked by referees [Amen]. The same attitude applies to academic positions. I would wager that no young researcher would be willing to jeopardize his or her scientific career by writing an essay such as this."

"This situation is particularly worrisome because there are good reasons to think the big bang (BB) model is seriously flawed," Burbidge continued. "One sign that something is amiss is the time-scale problem. The most favored version of the big bang model yields a universe that is between 7 and 13 billion years old. The large range of possible ages derives from uncertainty regarding the rate at which the universe is expanding, a value known as the Hubble constant."

"Comparison between observation and calculations of stellar evolution implies that the oldest known stars are 13 to 15 billion years old, with an uncertainty of plus or minus 20 percent. The estimated age of the elements in the solar system based on measurements of heavy radioactive elements is about 15 billion years, again including some uncertainty. If one accepts a high value of the Hubble constant, and hence a low-age universe, the simplest big bang model clearly fails, because the universe cannot be younger than the objects it contains. If one chooses a low value for the Hubble constant, it is touch and go," Burbidge concluded.

Recently, the Hubble Space Telescope (HST) provided new data used in revising the Hubble constant to 81-85 km/sec/Mpc. This figure proved disconcerting to advocates of the BB, because any result over 50 km/sec/Mpc reveals a serious flaw in the standard model.

Further, there is an inescapable problem of creation in both the BB and the steady state cosmologies with no scientific solution. How did the singularity or primeval atom originate? Could all matter of the Universe really have been packaged in such a small unit and at trillions of degrees temperature? If an explosive expansion was nearly instantaneous, why is the Universe now expanding at merely the speed of light? Why wasn't gravity an impending factor during the time of instantaneous expansion? With the Universe expanding at or near the speed of light (as extrapolation of the factors of the Hubble constant indicate), why are extrapolations backward to the very beginning done on this basis instead of the basis of instantaneous expansion (an important basis of the BB theory)? Hasn't the Universe always expanded at the speed of light?

Quoting Burbidge again, "Rather than consider alternatives to the big bang, cosmologists contort themselves and propose that the rate of expansion is just small enough to accommodate the oldest well-documented stellar ages. Alternatively, they vary the big bang model by invoking an arbitrary parameter called the cosmological constant. In this version of the story, the initial big bang was followed by a waiting period and then a further expansion." In the LB alternative, no cosmological constant is needed to explain the accelerating rate of expansion of the universe.

The Cosmic Microwave Background: The 2.7 K radiation

The second fundamental basis of the BB theory deals with the 2.7 K radiation identified as a relic of the big event. While this interpretation makes a good story, it does raise serious questions that leave the door open to other possibilities. But first, a little background would be helpful.

In 1965, two young scientists, Arno Penzias and Robert Wilson, decided to use sensitive microwave antenna in radio astronomy. Much to their chagrin, they discovered an irremovable background noise in the antenna. Frustrated, they sought help at Princeton University, where they were informed that the persistent "noise" was probably the most important radio signal ever to be received from outer space. This 2.7 K radiation, identified as the afterglow of the Big Bang, suffuses the sky in all directions at microwave frequencies.

The cosmic microwave background (CMB) of 2.7 K radiation was interpreted as providing direct evidence of how radiation was distributed throughout the Universe when it was less than one million years old. In 1989, the first Cosmic Background Explorer (COBE) brought back recordings of uniformity not varying more than 1 part in 100,000 in all directions. Advocates of the BB theory immediately interpreted these results, along with the thermal nature of the cosmic microwave background as evidence of its primordial origin.

In the same 1992 article, Burbidge wrote, "The pervasive cosmic microwave background was predicted by the big bang theory, and is still considered to be one of its strongest pieces of supportive evidence. Measurements now, however, show that the background radiation is extremely smoothly distributed. Maps of galaxies, on the other hand, show structure on all scales."

"According to the standard version of the big bang theory, matter and radiation were strongly coupled together in the early universe, and only later did the two go their separate ways. If this were so, the cosmic microwave background would show some imprint from the lumpy matter distribution that led to the formation of galaxies. In actuality, however, the cosmic microwave background appears smooth to at least one part in 100,000, so close to the level at which the big bang must be abandoned or significantly modified."

However, not to worry: The COBE Satellite had been sent aloft again, and recorded another cosmic microwave background of 2.7 K radiation. Just two months after the Burbidge publication, the interpretation of this COBE probe became public. Incredulously, the presence of miniscule ripples – small temperature fluctuations of slightly more than 1 part in 1 million – was announced, and advocates of the BB breathed a huge sigh of relief. The tiny ripples were interpreted as fluctuations in the density of matter and energy in a very early phase of cosmic history: the clumping of matter into larger structures such as galaxies.

In the same article, Burbidge had written, "Within the framework of the hot big bang, there is no satisfactory theory of how galaxies and larger structures formed. Galaxies cannot form by gravitational collapse in an expanding universe unless one assumes, without explanation, that large density fluctuations were present in the early universe. Under the influence of particle physicists, cosmologists are now proposing that these fluctuations occurred at an early stage of the big bang or else were caused by exotic entities such as cosmic strings. None of these ideas can be directly tested."

"The inflationary model, a pet idea of the past decade, holds that a period of extremely rapid expansion in the early universe accounts for both the smoothness of the cosmic microwave background and for the amount of matter present in the universe. But again, inflation is an untestable addition to the lore of the big bang."

Further, if "a period of extremely rapid expansion in the early universe" did happen, wouldn't this disrupt calculations of extrapolations backward to the time of the BB? To quote Burbidge again, "This form of inflation is arbitrary, and our successors will wonder when it goes out of favor, as the history of science suggests it will, why it was so popular?"

Whether or not the new finding of miniscule fluctuations in the background radiation eliminated the necessity of the inflationary model is not clear in many minds. But when we take a closer look at the CMB radiation, the question becomes moot.

A Closer Look at the CM Background Radiation

While the cosmic microwave background is identified as a relic of the big event, the true source of the 2.7 K radiation remains debatable. The major problem with this pervasive radiation resides in its interpretation. In reality, it is blackbody radiation, and such radiation is characteristic of all Solar System objects. (Ref: *Encyclopedia of Physics*).

Consider the radiation in the space enclosed within a hollow steel ball, which is maintained at a constant white heat. The radiation within such a sphere is called "blackbody" radiation because the radiation coming from each unit area of the walls of the enclosure is the same as that which comes from unit area of a perfect blackbody maintained at the same temperature. The radiation from each unit area of the walls of the enclosure is determined only by the temperature of the walls and not by the substance of which the walls are composed. In such an enclosure, the intensity of the radiation falling on a unit area of the walls equals the radiation coming from the unit area.

In principle, the same thing applies to a planetary sphere in which the white heat equates to its nuclear energy core and the mantle/crust equates to the walls. The radiated energy originates from the internal energy associated with atomic and molecular motion and the accompanying accelerations of electrical charges within the object (sphere).

Scientists realized that the results of the discovery of Penzias and Wilson were limited to only a few wavelengths clustered at one end of the Planck curve. Other explanations of the background radiation, such as a combination of radio sources, could explain those data points. It was not until the mid 1970s that enough measurements at different frequencies had been made to prove that the background radiation actually follows Planck's law.

Radiation is both absorbed and emitted by the walls of the enclosure, and by any gas, which might be in the enclosure. Planck suggested the idea that when radiation is emitted or absorbed, it is emitted or absorbed in multiples of a definite amount, identified as a quantum. The energy of a quantum of radiation of frequency v is proportional to v and so is equal to hv, where h is a proportionality constant known as Planck's constant.

The full curve of Planck's law is the spectrum of the continuous radiation given out by a hot blackbody. The plotted curve shows the distribution function with respect to wave-length and/or frequency. It applies to any planetary mass containing a hot nuclear core.

No problem would exist here if scientists could accept the fact that the spheres of the SS are evolving from nuclear masses, and the CBR is in reality the radiation emanating from these internal nuclear reactions. Simply put, the relationship among the three characteristics of these internal nuclear reactions is best shown by the full curve of Planck's law: the spectrum of the continuous radiation given out by these hot black bodies.

Proponents of the BB theory claim evidence that the CBR originated at extremely high red shift. The evidence comes from the production of light elements. Knowing the present CBR temperature, and assuming entropy was nearly conserved back to high red shifts, they trace the thermal history of the Universe back in time to temperatures high enough to have driven thermonuclear reactions. Computations of the nuclear reactions, under three other (conditional) assumptions, predict that most matter comes out of the "hot BB" as hydrogen, with about 20% by mass helium, and significant amounts of deuterium, ^{3}He and ^{7}lithium. The computation, when using a mean baryon number and consistent with what is observed [supposedly] to be present in and around galaxies, yields computed abundances concordant with what is seen in old stars.

Utilizing the same information except for the assumptions necessary in the BB theory, similar reasoning in the perspective of the Little Bangs (LB)/FLINE concept allows one to reach another confirmatory conclusion. Rather than looking to the BB for answers to the origin of light elements, the computation of the nuclear reactions can be applied most assuredly to the nucleosynthesis common to all planets and stars. No complicated assumptions are necessary. As shown in previous chapters, the production of all elements, therein, will be in the same proportions predicted in the calculations used in the BB perspective.

Planck's law applies to all thermonuclear reactions, no matter in which theory it is utilized. The above information ranks high among the many crucial clues that reveal the true nature of planetary origins. While negating any claim that the CBR exclusively supports the BB theory, the facts stand firmly as powerful confirmatory evidence of the blackbody nature of the spheres of our SS – the true source of the cosmic background radiation.

Consequently, any COBE search within the realm of our SS cannot escape the effects of the blackbody radiation emanating from objects comprising the system. Isn't it reasonable, then, to conclude that no matter where Penzias and Wilson pointed their microwave horn antenna, they inevitably would record the same noisy emissions in every direction?

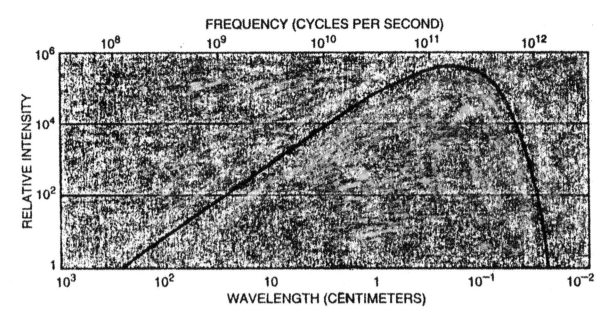

Planck's curve of blackbody spectrum applies to all thermonuclear reactions. It is powerful evidence of the blackbody radiation emanating from spheres of our Solar System–the true source of the cosmic background radiation recorded in every direction.
Figure 14

Even a probe escaping the effects of our SS would never be able to escape the "noise" of other suns (stars) undergoing transformation in the normal course of evolution. Starting with our own planet, it can be heard as the "noise" of nucleosynthesis – the hum of creation – that does indeed pervade throughout the Universe. As long as there is an active, expanding Universe, the hum will not cease.

Certainly, the eternal *Hum of Creation* is a suitable name for this 2.7 K blackbody radiation. Rather than a remnant of the BB, it is the humming of ongoing creation by which our planets evolve through the five stages of evolution and our Sun slowly sacrifices its mass in making the system possible.

However, keep in mind that this persistent hum of the spheres bears little relevancy to Kepler's long-sought, elusive music of the spheres. That tune dwells within the beautiful Phi geometry of the SS.

Misinterpretation of the significance of the 2.7 K radiation is a prime example of many discoveries that have been force-fitted into the BB theory in efforts to advance the cause of a concept already fatally flawed. An excerpt from a letter published in *Science News* (27 Jul 1991) reads: "When the Big Bang proponents make assertions such as 'a whole bunch of observations that hang together', they overlook [misinterpret contradictory] observational facts that have been piling up for 25 [now 35] years, and that now have become overwhelming."

The Relative Abundance of Elements via the Big Bang (BB)

The seed for Big Bang cosmology was watered in the 1930s, after Edwin P. Hubble, an American astronomer, discovered that galaxies recede from one another and that the most distant ones recede at the greatest rates. Hubble's discovery of this ongoing expansion was interpreted to imply that the cosmos had once been concentrated as a very small unit that exploded into our universal system.

To make the system work, scientists had to start with a tiny unit of mass at a temperature measured in trillions of trillions degrees. The tiny unit exploded instantaneously, spreading in all directions as a hot cloud of energy. Nuclear physics provided the tools for modeling the synthesis of the elements from fundamental particles within this hot cauldron.

The initial quantitative theory was developed by George Gamow, a Russian-born physicist, who had built a reputation by explaining radioactive decay.

In the early 1930s, scientists realized that most stars consist of hydrogen and helium. Since the hydrogen nucleus contains only one proton, it seemed reasonable to assume that it was the first element to form, or precipitate, from the hot expansion as it began cooling down. Helium, consisting of a nucleus containing four protons and four neutrons, was next in the line of elemental formation, created by the fusion of hydrogen.

However, the fusion of protons requires powerful forces to overcome the immense electrostatic repulsion between them. Scientists believed that the tremendous heat and pressure required in this process could be furnished only by a primordial event or the interior of a star. The prevailing theory of the nuclear physics of stars has been established by Hans Bethe in his explanation of how the Sun shines: nuclear fusion in stellar interiors converts mass into energy (and vise versa). This led to Hoyle's solution in which carbon is produced from three helium nuclei under the severe conditions in star cores.

By 1957, a scheme explaining how stars might have synthesized most of the elements from hydrogen and helium had been worked out by Fowler, Hoyle, Margaret Burbidge and Geoffrey Burbidge. A.G.W. Cameron in Canada did the work independently.

But the cosmic abundance of helium remained a mystery, which was eventually solved by Gamow. He suggested that the elements might have formed in an extremely hot, dense gas of neutrons, even before the stars came into existence. Some of the neutrons decayed into protons and electrons – the building blocks of hydrogen. According to this theory, larger nuclei formed in the primeval inferno when smaller ones, beginning with hydrogen, grew through the successive capture of neutrons. This process continued until the supply of neutrons was exhausted, the temperature dropped and the particles dispersed. Hoyle has the distinction of calling this the big bang theory, a concept of creation of the elements that prevails today.

Initially, the BB theory failed to explain the formation of the elements beyond helium, which has a mass number of four. Because there are no stable isotopes having mass numbers of five and eight, elements heavier than helium cannot be made by adding neutrons to it one at a time. Only invoking the stellar nucleosynthesis of Hoyle, Fowler and their associates solved this problem. Therefore, the heavier elements had to have been made in stars and supernovae.

Thus, the origin and the abundance of the elements found throughout the Universe are supposedly explained by the BB theory.

In a strange twist of fate, the BB theory and the works of Hoyle *et al.* have proven to be important steps in confirming the evolutionary process in which elements are also made in planets via nucleosynthesis. As evidence mounts against the BB theory, it grows more overwhelmingly in favor of the Little Bangs theory of the late 1970s. While it is true that elements are made via nucleosynthesis in stars, our Sun and in supernovae explosions, scientists have somehow managed to overlook an obvious source of creation of elements: nucleosynthesis in the SS spheres and other planets, moons, etc.

Here the BB can take much of the blame, which in turn, can be passed back to the Kant-LaPlace era in which the hypothesis of the formation of our Solar System from a gaseous dust-cloud originated. Scientists conveniently found a source of this material in the BB theory. This, in turn, led to erroneous conclusions that all universal spheres formed via condensation processes. Evidence shows that nothing could be further from the truth.

Quoting Burbidge again: "Big bang cosmology is probably as widely believed as has been any theory of the universe in the history of Western civilization. It rests, however, on many untested, and in some cases untestable, assumptions. Indeed, big bang cosmology has become a bandwagon of thought that reflects faith as much as objective truth."

Burbidge closes with this mutual thought: "Why, then, has the big bang become so deeply entrenched in modern thought? Everything evolves as a function of time except for the laws of physics. Hence, there are two immutables: the act of creation and the laws of physics, which spring forth fully fashioned from that act. The big bang ultimately reflects some cosmologists' search for creation and for a beginning. That search properly lies in the realm of metaphysics, not science."

The Fiery Little Bangs Theory: A Plausible Alternative

In 1993, the author found a copy of *Einstein's Universe* (1979) in which Nigel Calder wrote an opinion on the BB theory: "It may be that no one has yet thought of a better and truer scheme for the Universe." At that time, Calder was among the vast majority who were unaware that a more logical scheme had begun to take shape in the author's mind in 1973 and was being stubbornly pursued, persistently developed, and to a very limited degree, recorded in the scientific literature. The Little Bangs (LB) theory of 1979 became a vital link in the FLINE concept that finally came to full fruition in 1995 when the last Diagrams of the SS evolved into the Fourth Law of Planetary Motion as the final piece of the giant puzzle of our origins.

Even at this writing, new discoveries continue to add more and more exciting evidence, each bit driving another nail into the coffin of the BB theory while adding substantial support to the LB concept that ties everything together. Calder's book seemed to vindicate my findings of the previous 20 years, while giving incentive to push onward.

The initial clue to the LB was rooted in Hubble's discovery that galaxies are moving rapidly away from us and from each other, and the farther away they are, the faster they are moving. The BB interpretation that an explosive force imparted such vast momentum to the matter now comprising the distant galaxies did not make sense, especially since they were, and still are, moving aggressively, after billions of years, against strong forces of a centralized gravity.

Further, astronomers now know that galaxies at great distances are significantly younger than nearby galaxies. And on balance, present evidence favors never-ending expansion (not possible in the BB version) and the dire necessity of a continuity in nature. The youthful ages of distant galaxies furnish that continuity.

In my mind, the utilization of powerful forces at the spherical perimeter of the expanding Universe was essential in keeping the system functioning in a perpetual motion manner. Black holes and quasars could be the sources of the powerful energy so essential to the system. The concept fell together in 1979.

As the white energy of light reaches the spherical Universe at its perimeter, it is slowed somewhat, and collapses into the form of tiny black holes, which continue onward at nearly the speed of light while continuing to gather energy also from the darkness of the space it enters. A black hole eventually becomes a quasar, shining brilliantly because of the subsequent in-falling light crashing violently and noisily onto its perimeter.

It seemed reasonable to believe that the speed of the in-falling light increases dramatically as it is pulled into the new energy mass — perhaps attaining a velocity as high as the c^2 speed figure in Einstein's famous formula, $E = mc^2$. Perhaps this formula is telling us that energy is created by the in-falling mass traveling at the speed of light squared.

In the same book, Calder wrote: "Physicists working with atomic particles found that light of sufficient energy — very energetic 'gamma-rays' to be precise — could make fresh atomic particles. The energy of light was transformed into matter." At the time, I was unaware of both Calder's Book and the matter-antimatter reaction. By sheer coincidence, we had reached essentially the same conclusion through similar processes of thought during the same year. The LBT formulated at that time predicted that light energy existing at the perimeter of the spherical Universe coalesces into 'black hole' energy that later forms into bright, noisy quasars that eventually erupt into a series of galaxies, thereby explaining the speed-of-light expansion of the Universe (galaxies and quasars moving at that speed at the perimeter).

Calder continued the coincidence: "When a black hole is continuously fed with matter, its surroundings can glow extremely brightly. The falling matter exudes the energy, as if in a dying shriek, before it disappears forever. ...Quasars are the small cores of violently exploding galaxies, far outshining the ordinary stars."

This description was very similar to the LBT description of a quasar in describing its formation from a black hole, and later exploding into a series of galaxies – written the same year, but only discovered 14 years later. It served as a vital part of the LB theory. The manuscript was copyrighted, but unfortunately, no editor accepted it for publication. It remains a good dust collector.

Quoting Calder again "...when a particle meets its anti-particle, they annihilate one another and disappear from the Universe. All that remains is a 'puff' of gamma-rays – a reversal of the process of creation."

Further, "All that is required to create a black hole... is that light should feel the effects of gravity... Space cannot exist separately from 'what fills space', and the geometry of space is determined by the matter it contains."

At this time, it would be worthwhile to consider another, but similar opinion. Edward Tyron, City University of New York, has boldly stated that the Universe could have been created out of absolutely nothing without violating any of our physical laws, if it has certain properties. He wrote, "Every phenomenon that could happen in principle actually does happen occasionally in practice, on a statistically random basis. For example, quantum electrodynamics reveals that an electron, positron and photon occasionally emerge spontaneously from a perfect vacuum. When this happens, the three particles exist for a brief time, and then annihilate each other, leaving no trace behind. ...[This] is called a vacuum fluctuation, and is utterly commonplace in quantum field theory."

Continuing with the LBT: after ages of time, the brilliant quasar explodes, sending brilliant fireballs of energy embedded in clouds of energy-matter that spread in whirling finger-like formations. Out of this fiery soup, a group of galaxies is created in spherical bubble-shaped formation around the central point of the explosion. Such bubble-shaped formations are observable in the distant sky.

Closer examination shows a clear hierarchy of galaxies grouped into clusters and clusters into super clusters. Such formations do not seem possible in the time frame and perspective of the BB theory – in spite of the tiny fluctuations in the COBE recording.

Within each whirling galaxy, the fireballs react with cooler enveloping clouds. Fiery masses near enough to each other interact to form binary star systems, a common sight to astronomers. When some of these systems meet the right conditions of relative sizes, velocities and distances apart, the result can be a solar system like the three recently discovered: 51 Pegasi, 70 Virginis and 47 Ursae Majoris. In very rare cases, they go a step further and meet the special prerequisites that cause the smaller, faster mass to break up at the points dictated by Nature's Phi geometry (Chapter I). The result is a SS like ours.

In his excellent article, *Monsters at the Heart of Galaxy Formation* (*Science*, 1 September 2000), John Kormendy expounds on the history and most recent findings concerning the relationship between black holes and galaxies as viewed from the current perspective of the Big Bang (BB) concept of origins. He states that massive black holes "were first invoked in the 1960s to explain the enormous energy output of active galactic nuclei (AGNs) such as quasars. These super massive black holes (BHs) stand in sharp contrast to ordinary BHs, which have masses of only a few solar masses and which are well known to form when massive stars die. The origin of super massive BHs is unknown, and their existence long remained a hypothesis. By the mid-1980s, BH 'engines' had become part of the theoretical framework for understanding AGN activity, but evidence for their existence was still lacking."

"Surveys with the Hubble Space Telescope (HST) are finding BHs in every galaxy that has an elliptical-galaxy-like 'bulge' component. These observations...indicate that BH growth and galaxy formation are closely linked. These results have profoundly changed astronomers' views of BHs. ...BHs are becoming an integral part of our understanding of galaxy formation."

The recognition of this BHs/galaxies connection is a giant step in the direction of the Little Bangs (LB) concept (1979-1980) of the origin and ongoing creation of the Universe. In this revolutionary concept, BHs are created as the densest form of energy at the spherical perimeter of a Universe expanding at the speed of light. Eventually, the massive BH is triggered explosively into a small series of fragments in a bubble-shaped formation around the center of the system. Each fragment of the shattered BH spews its energy/matter from its two opposite poles, thus giving the pinwheel effect commonly seen in galaxies. In this LB perspective, BHs grow first, then later spawn the galaxies consisting of gases, dust, and the fiery masses of energy we call stars, all moving at great speeds. This scenario provides the ideal situation for dynamic interactions of these energy masses which can, and do usually, form into various combinations of solar systems, both binary and multi-planetary. The smaller masses eventually evolve into the planets and exoplanets now being discovered at an ever faster pace. Since 1980, this LB scenario has offered a viable alternative to the BB and its basic principle of accretion of universal spheres; such spheres always begin as fiery masses of energy and slowly evolve via nucleosynthesis into the spheres we see today.

Thus, the basic principle of the galaxies/BHs relationship described in the LB concept of 1980 is being confirmed by these recent discoveries. But rather than viewing their findings in the perspective of the LB, astronomers still attempt to interpret them in accord with the BB. This raises questions concerning the nucleosynthesis/evolution relationship. In Nature, everything evolves. Since the evolution of any sphere is not possible without the internal nucleosynthesis that drives it forward, how is it possible that nucleosynthesis of all the elements occurred during the first moments of the BB rather than in the individual spheres now undergoing evolution? The answers to this question remove any doubt about the fallacies of the BB/Accretion concept, while offering powerful substantive evidence for the definitive LB/FLINE concept. Conclusive evidence will stem from future discoveries about the true nature and sources of comets and asteroids during the first decade of the 21st century.

More on Galaxies

Among the recent findings of the Space Telescope was a distant elliptical galaxy billions of light years away that looked too modern at a time when the Universe was supposedly a tenth of its present age. Hubble images show that the spiral shape of galaxies is far more common among these youthful galaxies than in ones closer to us in space and time. Many astronomers had postulated that ellipticals would form more slowly than spirals. And all of them had formed so much earlier than astronomers expected galaxies of any kind to form.

Yet, going back 12 billion years ago, nine-tenths of the way back to the BB, Hubble had photographed a fully formed elliptical galaxy. When Duccio Macchetto of the Space Telescope Science Institute in Chile aimed the Space Telescope at this mass in the sky, he saw the unmistakable image of the elliptical object.

Proponents of the BB do not believe that the presence of the full-fledged elliptical so soon after the Universe is supposed to have formed casts doubt on the reality of the BB, but it does raise questions about the true nature of the beginning. One favored variant of the BB assumes the spawning of a Universe containing a density of matter high enough to eventually halt its expansion. But that model also predicts that primordial matter was distributed so evenly that galaxy formation was delayed. Consequently, Machetto's discovery may force cosmologists into a version of a less dense, ever-expanding, open universe, in which galaxies could take shape earlier. If so, it will be another giant step in the direction of the LB concept.

In *Asimov on Astronomy* (1974), Asimov made the following prediction: "In fact, even in a finite Universe, with a radius of 12 billion light years, there might still be an infinite number of galaxies, almost all of them (paper-thin) existing in the outermost few miles of the Universe-sphere."

This prediction might not be too far off. The photograph made by the HST in December, 1995 is the deepest archaeological dig in the history of galaxies. The great abundance of faint galaxies some five-sixths of the way to the alleged edge of the visible Universe presents a problem. Much interest will focus on the 1500 to 2000 bluish dots, which many astronomers believe are young galaxies in the distant Universe. The implications extend well beyond one tiny patch of sky. Based on it, a survey of the entire sky to the same depth would reveal a total of 50 billion faint objects. To make such a map would require a million years.

At these great distances, most astronomers expected galaxy numbers to decrease. The fact that they don't decrease expands the population of known galaxies and leaves an uncomfortably short time for them to form after the BB. Further, radio galaxies and quasars are more densely packed at distant phases of the Universe than near at hand. Here again, the evidence mounts against the BB theory, while adding powerful support to the LB theory of expansion and growth of the Universe at its spherical perimeter. Not surprisingly, all things in nature grow in similar manner; e.g., a tree grows upward and outward by adding new matter to the ends of all the branches and trunk.

The immensity of the known Universe continues to grow by leaps and bounds. These vast quantities begin to cast more doubt that they all could have been stored inside a pinhead. Simultaneously, they add significant credibility to the LB theory in which they are created continuously.

An interesting comment on the BB theory was made recently in the weekly *Ask Marilyn* column. Responding to an inquiry, she wrote: "I think that if it had been a religion that first maintained the notion that all matter in the entire universe had once been contained in an area smaller than the point of a pen, scientists probably would have laughed at the idea."

In expressing his cynical view of science, Planck also gave hope for the future. Writing in his *Scientific Autobiography and Other Papers* (1949), the great physicist argued, "A new scientific truth does not triumph by convincing its opponents and making them see the light, but rather because its opponents eventually die, and a new generation grows up that is familiar with it."

In the article, *Is There a Super Way to Make Black Holes?* (*Science News*, Sep 11, 1999) scientists discuss theories and new findings on the origins of black holes. In one theory, massive stars succumb to gravity after burning up their nuclear fuel, and collapse under their own weight to become black holes.

In the second model, the star explodes as a supernova, hurling its outer layers into space and leaving behind a dense, burned-out remnant called a neutron star. Debris from the explosion then falls back onto the neutron star, turning it into a black hole. Recent observations of a star that orbits a suspected black hole is offered as support of the supernova model of black hole formation (Sep 9, 1999, *Nature*). The research team found that the star's outer layers contain oxygen, magnesium, silicon, and sulfur in abundances 6 to 10 times those found in the sun. They were puzzled because of the belief that the light-weight star, only twice the mass of our sun, would never have reached an internal temperature high enough (believed to be greater than 3 billion kelvins) to forge high concentrations of these elements. They believe that the star's massive companion could have generated them by exploding as a super-nova, ejecting the elements into space before collapsing into a black hole. Scientists were in general agreement that this is the only way that could have enhanced the four elements.

These findings, assumptions and conclusions raise issues and questions. Ranging between atomic numbers, 8 to 16, on a scale of 92 for uranium, these four elements are lightweight and relatively easy to forge in the internal nucleosynthesis of all active spheres of the universe under specific combinations of high temperature and pressure that determine the types and quantities of each element produced. If, according to prevailing beliefs, these lightweight elements were formed solely in the Big Bang (BB), could they have been forged in the star's massive companion before the supernova explosion? Isn't this assumption actually confirming the internal nucleosynthesis/evolution phases of the FLINE model of planetary origins while presenting a good argument against the BB? Why would not such explosions create significant quantities of the heavier elements credited as being created in super explosions? For example, elements 99 and 100 are produced in explosions of hydrogen bombs. At best, these findings do add strong evidence to the FLINE model of creation of atoms via internal nucleosynthesis under extreme conditions of temperature and pressure, while raising genuine doubts about the creation of all lightweight atoms (and as originally believed, all heavier elements) in the BB.

At this time, we cannot reasonably conclude that black holes can be made only in the manners described in the article. Open minds can lead to other possibilities. We must question what happens at the ever-expanding perimeter of the universe when its exiting energies collide with the absolute zero temperature where all atomic motion ceases. Could these energies collapse into the densest form of energy: black holes that grow rapidly into powerful gravity masses? Would such a mass later burst into spinning twin fountains spewing masses of fireballs in a spiral configuration we call a galaxy? This likely scenario was aptly named the Little Bangs Theory (LBT) in 1979, the year of its conception. Unlike the BB, the LBT answers more questions than it raises. It blends precisely with the FLINE model; as such, both concepts need to be explored in-depth by the scientific community. Scientists need to rethink prevailing beliefs about the BB, solar system origins and planetary evolution. Determining the precise combinations of temperature and pressure necessary to create each of the atomic elements via internal nucleosynthesis within each active sphere during each stage of its evolution is a good starting point.

How Stars Form

On April 1, 1995 the Hubble Space Telescope snapped a photograph showing about 50 stars inside monstrous columns of dust and dense molecular hydrogen gas. The gaseous towers, six trillion miles long, resemble stalagmites rising from a cavern floor. At their edges can be seen finger-like protrusions, each with tips larger than our SS, in which the stars are embedded.

The information released to the public six months later interpreted the photograph in the perspective of the BB theory: the stars supposedly are created when "the gas collapses under its own gravity." As they grow, the massive stars produce huge amounts of intense ultraviolet radiation that hollows out a cavity around them by heating their surroundings, thereby making them visible. As the cloud gets boiled away, it uncovers stars buried there.

Interpreted in the LB perspective, the monstrous columns of gas and fiery stars are products of a tremendous explosion (or a series of explosions) that imparts the momentum essential for interactions that can result in active binary systems while scattering them eventually over the six-trillion-mile distance. The stars were created as fireballs in the explosion(s), rather than from the explosion's clouds in which many remain hidden from view. However, the two viewpoints do agree on the cavity formation aspect of the theory.

Cosmic Misfits Elude Star-Formation Theories (Science, 2 March 2001)

To quote from the subject article: "Astronomers have become increasingly perplexed over the last few years by a strange new class of celestial body. Too small to fit conventional definitions of brown dwarfs, they nonetheless move through star-forming regions in a manner that separates them from planets orbiting a star. Once seen as anomalies, their growing numbers are forcing astronomers to sit up and take notice. On 14 February, a Japanese team raised the stakes by reporting its discovery of more than 100 of these objects in a star-forming region known as S106." The objects do not neatly fit any conventional definitions.

"This poses a big challenge for the standard picture of star formation," says Shu-ichiro Inutsuka, a theorist at Kyoto University.

In addition to hundreds of brown dwarfs, the team spotted more than 100 fainter free-floating objects with estimated masses 5 to 10 times that of planet Jupiter. Their infra-red emissions placed the objects within the region. The discovery sheds new light on the ubiquity of isolated planetary-mass objects. One astronomer, Joan Najita, stated, "I think these kind of results show that the process that makes stars can also make things that are substellar."

Quoting again: "Most astrophysicists believe brown dwarfs and stars condense directly out of vast molecular clouds, whereas planets form in disks of matter swirling around nascent stars. Small lone bodies, however, don't mesh well with either scenario."

"Two theories about the origins of planetary objects shed light on the elusive creations, but fall short of supplying a complete answer. One proposes that they are ejected from young stellar systems, the other that they form from molecular cloud cores with masses too low to give birth to stars. But Inutsuka says neither idea can account for the large numbers of smaller objects spotted in S106."

The first proposal is very close to the teachings of the LB/FLINE model which predicted and explains the inevitable broad range of sizes of stellar bodies as ejecta initially from much larger masses of energy, perhaps quasars, black holes or large stars during the explosive birth of galaxies. Many of the ejected masses end up as free-floating singles, many as binary systems, and occasionally, but rarely, as a multiple-planet system like ours. In the chaos of such powerful energy systems, there seems to be no rules other than the Four Laws of Planetary Motion to govern sizes or combinations of interacting spheres.

Within the realm of the universal ongoing chaos of the LB/FLINE model in which all things appear to intermesh precisely, there seems to be no room or need to believe in a Big Bang, the primary cause of the perplexity of the astronomers.

The Irony of Iron in Stars and Planets

The news article, *Study reinforces Earth-not-alone theory* (*The Atlanta Constitution*, Feb 21, 2001) by Alexandra Witze, carried the subtitle, *Amount of iron in stars suggests planets like ours exist*.

To quote: "Rocky planets like Earth may orbit most of the 100 billion stars in the Milky Way, a new study suggests. Many astronomers have suspected planets may be common in the galaxy and that extraterrestrial life may have gotten started on other planets like Earth. But the new research increases the likelihood that Earth-like planets really do exist." The findings were announced by astronomer, Norman Murray at the annual meeting of the American Association for the Advancement of Science (AAAS).

At this writing, astronomers have discovered about 55 planets around stars other than our Sun; however, rather than rocky like Earth, they are huge and gaseous like Jupiter. Continuing the quote, "The new study's findings rely on a statistical analysis of how much iron stars contain, which [allegedly] is a potential measure of how much rocky debris has been incinerated in the stars' upper atmospheres. A lot of such debris could mean enough of it also existed to clump into planets."

"Murray's team scoured scientific studies for data on the chemical makeup of 466 stars. The analysis showed most of the stars contain an amount of iron equivalent to half Earth's mass. The iron probably got there as meteorites burned up in the star," Murray said. "And where there are meteorites, there are probably rocky planets. I can't point at a single star and say that one's got it [a planet]. But on average, the stars contain enough iron that they are probably accompanied by Earthlike planets."

It is unfortunate that researchers currently have no choice but to interpret their findings solely in the perspective of the erroneous Big Bang/Accretion concept. If viewed in accord with the teachings of the LB/FLINE paradigm, each and every nuclear energy mass continuously undergoes nucleosynthesis that produces its own atomic elements, especially the lighter third of our universal elements ranging from hydrogen and helium up through iron (and beyond). The 466 stars are no exception to this ubiquitous rule.

Their persistent self-sustaining production of elements negates the need to hypothesize erroneously that iron came from meteorites that burned up in the stars — speculation that would be hilarious if not for its tragic consequences.

The giant gaseous planets simply are relatively small masses of nuclear energy self-encapsulated with huge volumes of gaseous matter of their own making: the second stage of all planetary evolution. The most basic principle of the FLINE model is that nucleosynthesis and evolution are inseparable; one cannot exist without the other. Nucleosynthesis is the internal nuclear engine that drives the evolution of all active spheres universal. Planets do not depend on stars for their sources of iron; each of the two types of spheres creates its own iron. Thus, there is no factual basis for the conclusion that the amount of iron in stars suggests planets like ours exist.

The FLINE model both predicted and explains why such planets do exist, and makes it easy to understand definitively the origins and evolution of the countless billions of planets. The new Fourth Law of Planetary Motion assures us that Earthlike planets like ours do indeed exist, but in an extremely low ratio to the total number of planets in galaxies throughout the Universe.

Redshifts: A Shaky Measuring Rod for Astronomical Distances?

As reported in the article, *Radical Theory Takes a Test* (*Science*, 26 Jan 2001 p.579), nearly 3000 scientists and educators gathered in San Diego for the 197th biannual meeting of the American Astronomical Society, held in conjunction with the American Association of Physics Teachers during the second week in January, 2001. A radical theory presented by Margaret Burbidge, a noted astronomer, described how she and two collaborators — Halton Arp and Yaoquan Chu — had found a pair of quasars flanking a galaxy known as Arp 220. Redshift measurements reveal the galaxy to be only 250 million light-years away, while the redshifts of the quasars indicate that they are about 6 billion light-years away. This brought up questions of whether the finding is a chance alignment of two distant quasars flanking the closer galaxy, and whether redshifts are a valid and dependable method of measuring astronomical distances.

Most astronomers believe that redshifts are a result of the rapid outward motion of the quasars as the universe expands: the faster the outward motion, the greater the distance away, and thus, the greater the redshift. So far, Burbidge stated, 11 nearby active galaxies with high-redshift "paired quasars" have turned up since she discovered the first one 4 years ago. She believes that in each case, the paired quasars and the associated galaxy might be equally close to observers on Earth. This naturally brings up profound questions of the odds and significance of finding an X number of such paired systems among the billions of galaxies and quasars.

Quoting the article: "Arp goes further. He thinks the quasars originated inside the galaxies, as clumps of new matter created billions of years after the big bang. Arp and a handful of other cosmological dissidents believe that matter is still coming into being in some [most] parts of the universe, including the cores of active galaxies [and stars, planets and moons]. The newly created matter [usually] is flung out in two opposite directions, just like the radio-emitting jets of high-energy particles that stream out of many active galaxies [and spheres]. The high redshift of the ejected matter, they say, may be due to its youth (an idea developed by the Indian astrophysicist, Jayant Narlikar) or to relativistic effects."

The answers to the posed questions remain indeterminate at this time; however, we can look to the Little Bangs concept for some plausible explanations. As explained earlier, galaxies are created when fiery spheres (stars) of all sizes, along with voluminous gaseous dust-clouds, are forcibly ejected from the two poles of a spinning mass of energy, probably a quasar or a black hole, thereby creating the common pinwheel galaxy effect revolving around the remaining central energy core.

These huge, newly-created energy masses, continuously forming while moving outward at the speed of light in the vacuum at the ultra-cold perimeter of the universe before and during birthing of the galaxies, explain why astronomers observe these massive galaxies moving at an ever greater speed as they look ever farther into space. Naturally, the farther out they look, the younger the observable galaxies. This concept eliminates the need for a cosmological constant. Astronomers will never be able to observe the limits of the universe, simply because the light from these new systems cannot reach Earth before a vast number of ever newer and unobservable systems already have been created even farther out.

In the paradox of redshifts one must consider the beautiful correlation between the distance and age of stars — in accord with the Little Bangs theory. Since greater distances equate to younger systems, perhaps redshifts can be attributed to both distance and age. Quasars form near the perimeter of the expanding universe a relatively short time before ejecting their galactic systems, so it does not seem logical that the findings of Burbidge *et al.* present a sound argument against the validity of the redshift method for measuring astronomical distances. Their findings, at this time, do appear to be chance alignments of more distance quasars and closer-in galaxies. Or perhaps, as a good probability, the 12 closer galaxies in the paired systems were ejected into existence in the opposite direction of expansion and, thereby, slowed sufficiently to account for their apparently slowed rate of outward movement — as indicated by their redshifts — away from observers on Earth. Certainly, more evidence is needed.

Should We Believe the Big Bang Scenario? (Science, 8 December 2000}

This short essay reads as follows: The extrapolation by astrophysicists and cosmologists back to a stage when the universe had been expanding for a few seconds deserves to be taken as seriously as, for instance, what geologists or paleontologists tell us about the early history of our Earth: Their inferences are just as indirect and generally less quantitative.

Moreover, there are several discoveries that might have been made over the last 30 years which would have invalidated the big bang hypothesis and which have not been made — the big bang theory has lived dangerously for decades and survived. Here are some of those absent observations:

1. Astronomers might have found an object whose helium abundance was far below the amount predicted from the big bang — 23%. This would have been fatal, because extra helium made in stars can readily boost helium above its pregalactic abundance, but there seems no way of converting all the helium back into hydrogen.
2. The background radiation measured so accurately by the Cosmic Background Explorer satellite might have turned out to have a spectrum that differed from the expected "blackbody" or thermal form. What's more, the radiation temperature could have been so smooth over the whole sky that it was incompatible with the fluctuations needed to give rise to present-day structures like clusters of galaxies.
3. A stable nutrino might have been discovered to have a mass of 100 to 10 electron volts. This would have been fatal, because the hot early universe would have contained almost as many nutrinos as photons. If each nutrino weighed even a millionth as much as an atom, they would, in toto, contribute too much mass to the present universe — more, even, than could be hidden in dark matter. Experimental physicists have been trying hard to measure nutrino masses, but they are seemingly too small to be important contributors to dark matter.
4. The deuterium abundance could have been so high that it was inconsistent with big bang nucleosynthesis (or implied an unacceptably low baryon density).

The big bang theory's survival gives us confidence in extrapolating right back to the first few seconds of cosmic history and assuming that the laws of microphysics were the same then as now. (End of essay).

Responses: These 4 topics raise questions about the validity of assumptions made on the shoulders of previous assumptions. Each warrants an examination in the perspective of the LB/FLINE model.

1. In the arena of internal nucleosynthesis (IN) in the evolving spheres of the universe, scientists do not yet understand the parameters governing the type and quantities of any element produced in a specific time-frame during the cycles of any sphere's evolution; e.g., the extreme internal conditions responsible for the helium-to-hydrogen ratios in our sun, stars, planets, etc. Or why our small central star produces a number of additional light elements, but none heavier than iron.

    Larger stars can produce a number of the heavier elements, while Earth can produce the full spectrum of 92 elements, including the heaviest one: uranium. This tells us that Earth's heavily encapsulated energy core builds up far more severe conditions of extreme temperatures and pressures than our sun's interior (which is open to the severe cold of the vacuum of space). Since hydrogen and helium are the easiest elements to produce in spheres, how can one ever expect to find an object whose helium abundance is far below the amount predicted from the big bang' — 23% — or from the LB/FLINE model?

2. Since all actively evolving spheres are "blackbody" or thermal form, the cosmic background radiation (CBR) could not have turned out to have a spectrum that differed from the blackbody form. In view of the scattered positions of spheres, the radiation temperature could not have been perfectly smooth; its unevenness precisely meets the expectations of the LB/FLINE model. The CBR is the 2.7 K blackbody radiation emanating from active sources: galaxies, stars, planets, young moons, etc.

3. If a nutrino is found to be the smallest unit of energy, its mass should be the equivalent of 1 divided by c^2 (the speed of light squared). If so, they are indeed far too small to be important contributors to dark matter. Perhaps they are the most basic building block of all energy forms?

4. The low deuterium abundance is consistent with the LB/FLINE model's expectations, and is irrelevant to a big bang nucleosynthesis that never happened. Nucleosynthesis is ongoing, eternal — not a one-time big bang occurrence. It is the driving force that's both inseparable from, and crucial to, the evolution of active spheres and life itself.

Although these might-have-been discoveries that would have invalidated the big bang have not been made, they are of a speculative nature based on doubtful assumptions, and offer no positive evidence in support of the big bang. But many other discoveries do support definitive concepts that challenge its scientific validity while pointing clearly to a viable alternative. For example, recent discoveries of some 50+ extrasolar systems present enigmas to the big bang/accretion hypotheses, but present no problem to the LB/FLINE model of dynamic origins and evolution that both predicted and explains their mysteries.

When viewed from the LB/FLINE perspective, the four absent observations do not appear to justify pursuit of the big bang hypothesis at the expense of ignoring the newer concept that provides a well-substantiated foundation on which to accurately interpret all relevant scientific discoveries. One example: The accelerating expansion of the universe is the exact reverse of the slowing-down of the expansion as predicted by the big bang; however, it is an inherent principle of the LB/FLINE model, which predicted and explains this phenomenon without the need of a cosmological constant.

Should we believe the big bang/accretion scenario or the LB/FLINE alternative? Is the polka-dot beginning of the universe as misleading now as Ptolomy's antiquated Earth-centered universe was for 1400 years? Will future generations question how the big bang hypothesis (first recorded by Edgar A. Poe in 1848) could have survived so long in the face of so much powerful evidence against it, and no genuinely substantiated evidence to support it? Will they look to the Copernicus-Galileo experience and ponder why history was permitted to repeat itself? Time, as always, will render the best judgment.

Alexander A. Scarborough

Solutions to Anomalies: A Summary List

The ramifications of the geometric solution to the origin of the SS are immense. As the final phase of the FLINE model, it completes the continuity of evidence that weaves together all the anomalies of our SS. Most of the anomalies listed below have been discussed in this book or in previous writings. All are either solved or solvable in this new perspective.

1. Why planets obey Kepler's First Three Laws of Planetary Motion.
2. The complete geometry of the SS's origin.
3. The original and current spacing of planets.
4. The valid explanation of the enigmatic Bode's Law of the spacing of planets.
5. Why Neptune and Pluto do not obey Bode's Law.
6. Why planetary orbits are inclined to the ecliptic.
7. Why the displacement (AU) of each planet from its original orbit.
8. The highly elliptical and large inclination of the orbit of Pluto.
9. The speeds and directions of rotation and evolution of planets.
10. The slow spin of Venus.
11. The huge size of Jupiter (the Geometric Mean of the SS).
12. The tapered sizes of the planets.
13. How large moons formed and evolved contemporaneously with planets.
14. Why the Sun's equatorial mass has a 7° inclination to the ecliptic.
15. Why the Sun's equatorial mass rotates faster than the remaining mass.
16. The huge discrepancy in angular momentum of the Sun and the planets.
17. The nature of Earth's core: nuclear energy (ten clues).
18. The abiogenic origin and evolution of hydrocarbon fuels (1973-1980).
19. The origin of surface features: water, land, salt mines, etc.
20. The expanding Earth: increases in sea level and land.
21. How the Asteroids came into being: Olbers was right.
22. The craters on moons, asteroids, comets and planets: Mills was right.
23. How planets evolve through five common stages in accordance with size.
24. The physical and chemical differences among planets and moons.
25. Why species come and go: the extinction of the dinosaurs.
26. Why Earth and other planets are self-sustaining entities.
27. Plate tectonics: from highly mobile to barely movable.
28. Why the electromagnetic strength of each planet is different.
29. The unexpected powerful explosions of Comet SL9 on Jupiter.
30. The origin and evolution of comets.
31. The origin and fate of planetary rings.
32. The cometary moons of Mars: Phobos and Deimos.
33. How the discoveries of space probes fit into the FLINE concept.
34. The 2.7 K radiation throughout the Universe: not from the Big Bang.
35. The erroneous notions of the Big Bang.
36. The too-high Hubble constant revealed by the Hubble Space Telescope.
37. The new conflict between the oldest stars and a younger Universe.
38. How and why the Universe expands at its spherical perimeter – forever.
39. The 51 Pegasi system discovered by the HST.
40. The large extrasolar planets: 70 Virginis and 47 Ursae Majoris.
41. The origins of known extrasolar systems (50+) and binary systems.

The full significance of the Fourth Law of Planetary Motion yet lies camouflaged within the mysteries of the SS and among the enigmas of the Universe.

Einstein was dogmatic about continuity in Nature, and rightly so. Nigel Calder wrote, "The uncomprehending antagonism evoked by Einstein's idea is a sign that the old conflict between scientific inquiry and dogma is far from dead." Perhaps it always will be so.

Quoting Calder again: "…great syntheses are made in individual minds." [Not by committees, as has been shown many times in the past]. "…all that is lacking is new comprehensive insight… That must come soon."

To the credit of science, the majority of scientists have significant doubt about the Big Bang theory. However, as long as the concept prevails in the news media, any new insight will have a tough uphill battle just to establish a foothold with that majority. At this writing, the revolutionary LB/FLINE concept has not been permitted to establish that foothold, in spite of the many substantiated facts comprising its sturdy structure.

One day in September of 1992, in frustration of the situation, I wrote: "Even as the 21st century approaches, the science of planetary origins and evolution remains bogged in a quagmire of egoistic bias, snugly entrenched in an ivory tower built of cards on a foundation of sand, with shades drawn against the light of new ideas and factual knowledge, reveling in the fantasies of hypotheses that would burst like a bubble with the singular probing of the finger of question."

While it did relieve the frustration somewhat, the situation has shown no perceptible sign of change. Perhaps some day the system will improve, but right now, there is no light at the end of the tunnel.

Perhaps this ninth edition of forced self-publications will add enough substance to the already overwhelming evidence against current dogma to force a change in the direction of scientific thought on the origins of our SS and evolution of its planetary systems. In view of the continuity of evidence herein, certainly the LB theory is a plausible alternative to the BB theory. The great beauty of it is that no changes in established evidence should be necessary; however, it is essential, even crucial, to the advancement of the planetary sciences and astronomy that the perspective in which this evidence is interpreted be examined closely.

In a battle for truth, one cannot afford to tread too lightly – or too heavily.

More Signs Black Hole Exists in Milky Way Center. September 7, 2001

An interesting article on the black hole at the center of our Milky Way appeared in the journal, *Nature,* on September 6, 2001. Scientists using the powerful new Chandra X-ray telescope garnered evidence that apparently clinches the case for the existence of a supermassive black hole at the center of our galaxy. Through NASA's $1.5 billion telescope, astronomers observed a flare of X-ray energy produced where the lip of the invisible black hole should be.

The clear-cut image of the flare was the first of its kind; it dimmed and brightened over 10 minutes, the time it would take for light to travel about 93 million miles around the lip of a black hole. Scientists calculated the mass of the black hole to be 2.6 million times that of our Sun, all packed in a small mass.

"We are now able to say that indeed all of the mass, by implication, is within that small region, and there is nothing we know that can be that dense and not be a black hole," Frederick Baganoff said.

The black hole is 24,000 light-years from Earth. Scientists now believe there are billions of black holes throughout the Universe, varying greatly in mass and luminosity.

More good news came that same evening from TV's *Stephen Hawking's Universe*. One astronomer stated that most scientists now agree that the Universe will expand forever. Thus on the same day, two stunning releases provided powerful support of the Little Bangs Theory in which black holes endlessly form at the perimeter of the Universe, moving initially at the speed of light, gradually slowing while spewing out galaxies of fiery masses and hot gaseous dust, thereby accounting for the observed accelerating expansion of the Universe in every direction, as well as for other anomalies of astronomy. The LBT, thus, quietly continues to provide answers to critical questions raised and unanswerable by the Big Bang hypothesis.

Einstein's general theory of relativity expands the time and space proposals of the special theory of relativity from the areas of electric and magnetic phenomena to all physical phenomena, with emphasis on gravity. Black holes, the densest form of electric, magnetic and gravity phenomena, are the medium through which energy is created at the perimeter of the Universe and distributed forcefully into galactic systems, each consisting of billions of brilliant stars enshrouded in gaseous dust-clouds. Our Milky Way is but one among billions of galaxies transformed in like manner, each from a black hole, a remnant of which normally remains at the galactic center. The new X-ray discovery is another confirmation of the LB concept in which our Universe is continually transformed into black energy destined to become brilliant white galaxies in which solar systems form primarily as binary systems and, very rarely, as multiple-planetary systems similar to ours.

In this manner does the LB/FLINE paradigm blend with the theories of relativity that expands the time and space proposals from the areas of electric and magnetic phenomena to all physical phenomena that eventually evolves into spheres we call planets, moons, comets, and gaseous dust-clouds. Asteroids, correctly identified by Olbers in 1802, are remnants of one or more disintegrated planet(s).

Letters and Memos

To: Glenn Strait, Science Editor, *The World & I*, Washington, DC. 01-19-98

Quoting from *A New Look at Black Holes* (*Science News*, Vol. 152, Nov 29 1997, pp.346-347): "Several researchers have recently invoked the advection model to explain the origin of a ubiquitous X-ray background that bathes the Universe..."

"Although advecting black holes don't emit much radiation, their feeble spectra do match the overall intensity pattern of the background radiation. If enough of these black holes exist, they could produce the observed X-ray emission, Tiziana Di Mateo and Fabian reported in the April 1 *Monthly Notices of the Royal Astronomical Society*. Insu Yi of the Institute for Advanced Study in Princeton, N.J., and Stephen Boughn of Haverford (Pa.) College present a similar argument in an unpublished article posted on the Internet."

As you might or might not be aware, my work on black holes began in 1979 and eventually led to an explanation of the source of the ubiquitous 2.7 K radiation currently thought to be a leftover from the Big Bang (BB). To elucidate, herewith are copies of pages 114-116 from my most recent book, *The Spacing of Planets: The Solution to a 400-Year Mystery* (1996), which finalize my previous work. You will note some similarities and connections among these writings; the main point is that all background radiation emanates from tangible sources rather than as a leftover from the BB.

The newly published evidence in *Science News* adds credibility to my writings on the origin of the ubiquitous 2.7 K radiation throughout the Universe. Knowing the true source of this weak radio signal negates one of the three observations that provide the fundamental basis for the standard cosmology featuring the BB theory. The other two observations include: (1) the observed expansion of the Universe (usually interpreted in the framework of general relativity as an expansion of the metric of space); and (2) the alleged successful explanation of the relative abundance of the light elements.

However, both of the latter observations can be explained more logically via the Little Bangs theory (LBT) (1979-1980) in which (1) the Universe is expanding and growing at the speed of light in all directions, and (2) the creation of elements was, and is, accomplished via internal nucleosynthesis within all active (nuclear) masses of the Universe (1973-1975), and is the source of the ubiquitous 2.7 K radiation. The LBT is in full compliance with Dirac's belief that our Universe began as a tiny 'singularity' (a single point), and that it will continue to expand forever. The LBT explains how this was, and is, accomplished.

Perhaps these recent writings, and others surely to follow, will open up the new perspectives of the proposed FL/IN paradigm — and make the inevitable death of the BB easier for scientists to accept. In follow-up, I will send a similar letter on comets: the new perspective vs. current beliefs.

Thanks again for your encouragement and understanding during this revolutionary transition in origins.

To: Glenn Strait, Science Editor, The World & I, Washington, DC. 02-09-98

I feel compelled to comment on four new writings in the scientific literature received this week. James Glanz's article, *Exploding Stars Point to a Universal Repulsive Force* (*Science*, 30 Jan 1998, pp. 651-652), states: "Not only did the results support the earlier evidence that the [cosmic] expansion rate has slowed too little for gravity ever to bring it to a stop; they also hinted that something is nudging the expansion along." Perlmutter concluded: "That would introduce important evidence that there is a cosmological constant." Michael Turner added, "What it means is that there is some form of energy we don't understand." The article continued, "Other observers had already found signs that *the universe contains far less mass than the mainstream theory of the big bang [BBT] predicts, which left open the possibility that some form of energy in empty space could be making up the deficit...* But the mass just doesn't seem to be there... Both teams concluded that *the expansion had slowed so little that it will probably go on forever.*" (*Science*, 31 October 1997, p. 799). While these findings present ever more problems for the BBT, they fit precisely into the Little Bangs Theory (LBT) of 1979-1980 — described in my earlier writings — in which a cosmological constant is not needed as a repulsive force to complete the big picture. Einstein was the first to propose a universal repulsive force, which he later — and according to the LBT paradigm — correctly abandoned.

The basic principles of the LBT in which the Universe expands at it spherical perimeter at the speed of (exiting) light (via energy/matter interactions) is further enhanced by another article: *Gamma Rays Create Matter Just by Plowing into Laser Light* (*Physics Today*, Feb 1998, pp. 17-18). To quote: "But the SLAC experiment was *the first direct observation of material particles produced by nothing but protons*," which "*can also be thought of as 'the sparking of the vacuum'*, an exotic prediction of quantum electrodynamics at extremely high electric field intensities." I believe Dirac initiated this sparking concept in the early 1920s, and other name scientists have added to the concept.

In *Comet Shower Hit, But Life Didn't Blink* (*Science*, 30 Jan 1998), "researchers tapping a new sort of geologic record find solid evidence that at least one comet shower did pelt Earth —but without apparent effects on life... the Massignano sediments yielded a 2-million-year-long surge in helium-3. It peaked about 35.5 million years ago, right at the time of the two major impacts" that allegedly caused the extinction of the dinosaurs. The evidence, per se, is in perfect harmony with the Energy Fuels Theory (EFT) of 1973 in which significant quantities of helium-3 show up with gas from deep-Earth drillings. The new EFT reveals that all elements comprising these deposits were (and still are) created via internal nucleosynthesis. Thomas Gold presented an excellent paper at the AAAS Meeting in 1985 on the gas/helium-3 association in deep-Earth drillings — which added strong evidence to the EFT. Unfortunately, an incorrect version of the 1973 EFT became known as Gold's theory in the 1980s. (See enclosed copy).

Next, *The New Gamma-Ray Astronomy* (*Physics Today*, Feb 1998, pp. 26-28) presents a good discussion on nucleosynthesis sites. "The modern era of theoretical nucleosynthesis sprang from a classic 1957 paper by Burbidge *et al.*" in which heavier elements are created in stars, supernovae and novae rather than being components of primordial gas from the big bang. The next logical step came from my realization in 1973 that planets evolve(d) from smaller masses of star-like energy — each one creating all elements comprising its atmosphere and/or crust via internal nucleosynthesis (IN) — until the nuclear core is depleted. Ironically, this concept inevitably led to the solution to the proposed Fourth Law of Planetary Motion (1980-1995) (now being refereed): the last phase that finally brought the longtime evolving FL/IN paradigm full circle. Now, as with the four papers discussed herein, all relevant articles and discoveries continually add more supportive evidence to this new perspective.

The fourth article, *The Origin of Chondrules at Jovian Resonances* (*Science*, 30 Jan 1998, p. 681) states: "Isotopic dating indicates that chondrules were produced a few million years after the solar nebula formed. This timing is incompatible with dynamical lifetimes of small particles in the nebula and short time scales for the formation of planetesimals." The authors present a tentative explanation of the puzzling discrepancies by attributing them to "Jovian resonances that cause collisional disruptions and melting of dust by bow shocks in the nebular gas" — an explanation based more on erroneous speculation than fact.

The basic problem here is the outmoded hypothesis that planetesimals accrete(d) from a nebula of dust and gas. The findings can be interpreted more accurately in the perspective of the FL/IN paradigm in which planetesimals are remnants of the [three] Asteroids planets that exploded — just as Olbers concluded in 1802. The proposed Fourth Law of Planetary Motion and Kepler's First Three Laws clearly reveal how the original Asteroids masses of energy attained their orbital positions around the Sun before beginning the five-stage evolution common to all planets. The ages of chondrules should indicate the time(s) the disruptions (perhaps explosions) occurred. Chemical analyses of their compositions should reveal much about both the internal and explosive nucleosynthesis aspects of these evolutionary events. For example, there are direct connections between the nickel-iron contents in some asteroids and meteorites and the results discussed in *The New Gamma-Ray Astronomy* article above. Finally, the findings described in the four articles can be tied neatly together via the FL/IN paradigm.

These four examples illustrate a crucial point: it is highly probable that interpretations in the perspective of the FLINE paradigm will provide correct answers to the many problems now existing throughout the prevailing beliefs about the origins and evolution of planetary systems. The same thing can be said for the LBT vs. the BBT. At the frontier of research, these new ideas seem destined to stimulate new directions for research; for example, in the basics of internal nucleosynthesis and its connections with El Nino and all other phenomena (on Earth, other planets and moons).

Unless someone is able to find a flaw in the LBT/FL/IN paradigm, my faith in its scientific validity remains steadfast. Thanks for taking the time to read this long letter, and for your other considerations.

P.S. - Most of the evidence in C.W. Hunt's *ANHYDRIDE THEORY: A Theory of Petroleum and Coal Generation* (1997), herewith, is in agreement with my *Energy Fuels Theory* (EFT) (1973). However, as with Thomas Gold (1980), Hunt misses the crucial starting point: (C & H molecules) methane created via internal nucleosynthesis, followed by polymerization and cross-linking into petroleum and coal.

P.P.S - Publication of the EFT and the LBT/FL/IN paradigm will lead to a provable understanding of the origins of solar systems and the evolution of all planetary systems.

To: Glenn Strait, Science Editor, *The World & I*, Washington, DC. 04-15-98

Thanks for your interesting Internet article of 98-04-10 concerning the vast amounts of water found in Orion. You asked how the explanation, therein, for the origin of water compares with my explanation. In the FL/IN/E paradigm, water (and all other planetary matter) are end products of internal nucleosynthesis (IN) that drives evolution (E) in all active spheres of the Universe. *IN and E are inseparable: one cannot exist without the other* (a fact that deftly eludes advocates of the Big Bang) (BB). *The Four Laws of Planetary Motion (FL) irrefutably support the fiery origin of solar systems and the evolution of the orbiting energy masses that gradually evolve into planetary matter through five common stages of evolution in accord with size and in full compliance with all natural laws.* This also explains the anomalies of exoplanets (e.g., most are too close to the parent star to have condensed from a gaseous dust-cloud and/or small planetesimals).

The Universal Law of Creation of Matter via IN applies equally to the Orion water — an end product of IN — expelled into space in a manner yet to be detailed.

With the headline that a "Factory found in Orion would fill Earth oceans 60 times per day," BB advocates appear to be stepping gingerly into the FLINE paradigm. Are they saying that water is still being created in cold space? and with or without internal nucleosynthesis? What is the source of the water's oxygen? I am under the impression that *all water (and all other matter) had been created during cool-down from the BB, and everything coalesces via gravity*. In any case, the interpretations expressed in the article appear to be far-fetched and force-fitted to the occasion, *while the aforementioned fact that IN and E are inseparable, and one cannot exist without the other*, is ignored.

Even though the BB admittedly is highly speculative and contains a number of fatal flaws, many advocates remain adamant in their refusal to consider alternatives — as was illustrated in the Internet article. The problem with their interpretations stems from this situation.

To: Dara Horn, Harvard University, Cambridge, MA 02138.   June 4, 1998

Congratulations on your excellent and perceptive article, *The Shoulders of Giants* (*Science*, 29 May 1998). During 63 years of study and research in various fields of science, I have never read a more tactfully perceptive description of the way things are in real life. However, I must admit my genuine surprise that such an exposé could be written by anyone so young, and that it could get past peer reviewers for publication in this magazine.

Cecilia Payne-Gaposchkin's fate is a sad commentary on science. Some years ago I read of a similar situation in *The Cry and the Covenant* (author, Thompson) in which Dr. Samuel Semmelweiss met an even sadder fate for his perceptive contribution of cleanliness during childbirth in the mid-to-late 19th century. If you have not read the book, I strongly recommend it to you for further understanding of real life in the sciences.

Many scientists have met, and still do meet, similar undeserving fates. My most familiar example is my present situation. For the past 25 years, my research in the field of planetary origins and evolution inevitably led to the proposed Fourth Law of Planetary Motion (1995) that first eluded Kepler (the discoverer of the First Three Laws) in 1595. The geometric solutions to the new Fourth Law (FL), together with Kepler's First Three Laws, detail how the planets attained their current spacing around the Sun — eventually to evolve via internal nucleosynthesis (IN) (from energy to atoms) into their present stages of evolution (E) — all in full accord with size. In our Solar System, the three facets of this FL/IN/E paradigm are inseparable (until the energy-source core is depleted): one is not possible without the other two. *Powerful supportive evidence continuously mounts; every relevant discovery serves as additional corroborative proof requiring no speculation, while leaving little, if any, room for doubt.*

Currently, the major problem is the one you described so well concerning Cecilia's situation. As with Copernicus, this revolutionary concept upsets prevailing speculative beliefs about solar systems and the evolution of planets. As such, it cannot get past peer reviewers schooled in current dogma. To preserve it for posterity, I was forced to *self-publish at great sacrifice to my budget. Grants, obviously, were unattainable. For the past three years, most of my abstracts and papers have been rejected without cause, reason or explanation by members of my science organizations — censorship at its finest.

The irony is that the total package of the FLINE paradigm explains why the stars are "amazingly uniform" in their composition and why "hydrogen is millions of times more abundant than any other element in the universe." Also, why and how light and heavy elements are created within Planet Earth.

Unfortunately, we both have been disillusioned: science is not the pure and objective pursuit of knowledge; rather, it has evolved into a political big-money game rift with the emotions best described in your perceptive article. The question now is what can be done by younger people like yourself to bring science back to where it should be. For the sake of science and posterity, I hope you will continue to use your tremendous talent as a writer always to strive towards this goal.

*THE SPACING OF PLANETS: The Solution to a 400-Year Mystery. (1996).

To: K.C. Cole, Journalist, Los Angeles Times, Los Angeles, CA 90053

Your excellent article headlined *33-year-old changing way astronomers see universe* (*Atlanta Journal-Constitution*, Sep 13, 1998) featured Andrea Ghez's presentation in August at Rutgers University in New Jersey. Her paper adds corroborative evidence to a revolutionary concept that has evolved during the past 25 years: the little-known FLINE paradigm of origins of the Universe and its solar systems and the orbital spacing and subsequent evolution of planets. The lack of familiarity with this valid scientific concept, perhaps, stems from its serious threat to current beliefs about such origins.

The FLINE paradigm explains why "a massive black hole sits at the center of the Milky Way" and at the center of all galaxies — and why a black hole is essential to the formation of each and every galaxy. In the perspective of the Little Bangs Theory (LBT) of 1980 — a vital part of the FLINE paradigm — the explosive black hole spews out, in pinwheel fashion, the dust, gases and fiery masses of energy we call stars. Obviously, such close initial proximity accounts for Ghez's "discovery that most newborn stars appear to be twins." Seldom can they remain single stars. Clearly, they usually were born double; however, some can drift apart later as single stars.

In similar manner since 1979, the evolving FLINE paradigm has contradicted the prevalent theories of how stars form from dust and gases — which, in reality, are byproducts created simultaneously with the fiery stars comprising the galaxy ever since their explosive ejections from the twin outlets of a central black hole. Obviously, this concept adds dramatically to the large amount of evidence already contradicting the Big Bang hypothesis of the origin of our Universe.

The new paradigm provides the perfect (and perhaps the only logical) mechanism to explain the formation of double stars, including the recently discovered exoplanets. Ghez's belief that the current theory "very nicely produces our sun and planets" continues to ignore the Four Laws of Planetary Motion (FL) detailing how the planets attained their orbital spacing around the Sun. The FL clearly reveal how our dynamic Solar System formed as a special and rare multi-star system via the same explosive forces that formed all binary star systems. Planetary evolution (E) is a fact of Nature, but such ongoing changes are not possible without a source of energy to drive them forward through the five stages of evolution common to all planets: from energy to gaseous to transitional to rocky to inactive (dead) spheres. In scientific terms, these processes are known as internal nucleosynthesis (IN) (the creation of atoms from energy, per $E = mc^2$) and polymerization (the combining of atoms to form all planetary matter). The complete details are found in my writings of the past 25 years.

In further confirming the FLINE paradigm, Ghez clearly reinforces this revolutionary concept to the extent that it now seems destined to displace prevailing beliefs about planetary and universal origins early in the 21st century. Perhaps her paper and the many corroborative findings of space probes will encourage scientists, finally, to give more serious consideration to the FLINE paradigm of 1973-1998. If you care to delve deeper into this new concept for more stories, I would be happy to work with you.

To: Glenn Strait, Science Editor, The World & I, Washington, DC.   09-03-98

Your e-mail on Electric Space does stimulate thinking. I'm convinced that there is a lot of truth in the plasma model of the Universe, and that it does offer significant challenges to the Big Bang (BB). However, it does fit well in the Little Bangs theory (LBT). As far back as science is able to comprehend, everything in the Universe begins with electric space. We may never know how this basic form of energy came into being; however, the LBT describes its continual creation out of empty space at the rapidly expanding spherical perimeter of our Universe.

Nature's next step is the transformation of "99.999 percent" of the basic electric energy (BEE) (which some scientists call aether) into other forms of energy (i.e., black holes and quasars) that form at the perimeter of the ever-expanding-at-the-speed-of-light Universe. From these sources spring the explosiveness that creates the galaxies of plasma comprising their stars, dust, gases, solar systems, etc. In this explosive mix, the observable plasma energy forms are usually misinterpreted in the perspective of the BB as stars created via condensation of the dust and gases. In reality, the starry fireballs of plasma energy are byproducts of the powerful explosions or outpourings of these forms of energy. Under the proper conditions, smaller masses become locked in orbit around larger masses in the initial phase of binary and —in rarer cases —multi-bodied 'solar systems'.

This revolutionary concept is corroborated strongly by the Four Laws of Planetary Motion — which are vital factors in explaining the anomalies of our Solar System and exoplanets. Once placed in orbits around a Sun, each fireball of plasma begins its transformations through the five stages of evolution common to all planets and moons — until the energy source is depleted (totally transformed into planetary matter). Evolution ceases, and everything eventually grinds to a halt at the end of the long journey of transformations from the original plasma energy into its ultimate destination: atomic matter.

The fact that "plasmas generate and react strongly to electromagnetic fields" clearly confirms the true nature of planetary cores, thereby providing convincing evidence that no other source (i.e., the geo-dynamo) is necessary to create planetary electromagnetism. The fact that "plasmas are also prodigious producers of electromagnetic radiation, from very low frequencies, to microwaves, to very high frequencies such as those associated with cosmic rays" is interpreted in the FLINE paradigm as an explanation of the 2.7 K radiation and the recently discovered infrared glow throughout the Universe, both of which have been identified erroneously as leftovers from the BB. The FLINE interpretation renders a better understanding of the significance of Planck's curve of blackbody spectrum that applies to all thermonuclear reactions —which further explains why the cosmic background radiation can be recorded so smoothly in every direction (as detailed in my latest book, *The Spacing of Planets*).

In confirming my writings on the subject, this press release brings researchers another giant step closer to the FLINE paradigm and to "a revision of our understanding physical processes in space as far ranging as the formation of planets to the sources of high energy particles and radiation."

To: Glenn Strait, Science Editor, The World & I, Washington, DC.   09-25-98

Thanks for the information and suggestions in your e-mail of 98-09-15 under the subject of *Common Sense Science*. The address you requested has been mailed in a previous response to another subject, but I would like to make some comments.

I feel at fault in not being able to convey the connection between the Little Bangs theory (LBT) and the FLINE paradigm for the origin of the Solar System and the evolution of planets. The link resides in the necessity of continual creation of energy and its subsequent explosive manner of conversion into lesser forms of energy systems such as galaxies of binary and multiple star systems of fiery masses that subsequently are transformed into the atomic form of matter. The LBT is to the FLINE paradigm as the Big Bang is to the accretion concept of planetary origins. The basic principle is the continual transformation of universal energy from one form to another in full accord with the prevailing conditions imposed upon it. But I will leave the cosmology details to others, as you suggested.

The FLINE paradigm has evolved slowly during the past 25 years of carefully piecing together as many relevant and factual bits of information as feasible from the scientific literature. The primary objective was to separate facts from myths. I feel that this guiding principle has forced me into meeting that goal. The really surprising thing is the huge number of facts that have accumulated during those years of research and how well they interlock throughout without the need for assumptions and speculations. The best analogy for me: the finished jigsaw puzzles of many years ago in which it was essential to interlock each piece into its correct position. If a piece did not fit into a position, it could never be force-fitted. When finished, all the pieces fitted precisely in place to reveal the big picture.

My weaker point does reside in the field of sub-atomic physics, even though I have gained some knowledge of it through the years. So I must leave the details of the internal nucleosynthesis of stars and planets to the better trained physicists. However, the finer details really are not crucial to the structure of any valid concept as long as there is sufficient evidence, otherwise, to corroborate its every facet. The vast quantity of such evidence corroborating the FLINE paradigm is truly surprising. Its IN/E facets forced me into finding the final piece of the puzzle that ties everything together: the solution to the Fourth Law of Planetary Motion (1980-1995) that first eluded Kepler in 1595.

As to the neutrino puzzle, my best guess at this time is that the answer might reside in the differences between the parameters of the two sources you mentioned: our Sun and planets. One is a huge, open system; the other is a very small closed system. My chemical engineering professors at Georgia Tech stressed that the differences in two similar systems will account for differences in the end results. This lesson proved to be crucial to understanding the differences in elemental compositions of planets, caused by the progressively changing conditions (e.g., encapsulation, pressure, temperature, concentration, etc.) within nuclear energy cores as functions of time and planetary size — which explains why Earth can produce elements up through uranium, while the Sun can produce elements only through iron.

Allow me to pose two questions: How does one distinguish between a neutrino from the Sun and one from our planet? Can neutrinos be detected and identified as emanating from Jupiter (in only the second stage of evolution)? Since I don't have the answers, any other suggestion of mine would be in the realm of pure speculation.

To: Christine Gilbert, Letters Editor, *SCIENCE*, Washington, D.C.   12-29-9

In his excellent article, *Cosmic Motion Revealed* (*Science*, 18 Dec 1998), James Glanz discusses how "astronomers peered deep into the universe and found that it is flying apart ever faster, suggesting that Einstein was right when he posited a mysterious energy [lambda] that fills 'empty' space."

Without lambda, this discovery conflicts with the Big Bang theory (BBT) in which just the opposite (a slowing down at these great distances) is more likely. Quoting Glanz: "Back in 1917, when Einstein proposed the [lambda] constant, he thought the universe was static, neither expanding nor collapsing. He put the cosmic repulsion [lambda] into his equations to prevent the universe from collapsing on itself from the gravitational pull of the matter inside it." Years later, Einstein "reasoned that if the expansion was a relic of a primeval explosion, the cosmological constant — which he felt made the equations unaesthetic — wasn't needed. He withdrew the idea and called it [correctly] his 'biggest blunder.'"

In the BBT, the observation that distant matter is flying apart ever faster must, of necessity, utilize a lambda factor. However, there are caveats. So far, "calculations suggest that such a lambda should be many orders of magnitude larger than the supernova groups have seen. That puzzle has launched a search for new physics principles... At this point the cosmological constant remains in the realm of theory; no one yet knows the precise nature of the energy causing the universe to fly apart ever faster." To exacerbate this odd situation, volumes of relevant substantiated evidence against the BBT continue to be ignored.

Back in 1980, a revolutionary concept of universal origins accurately predicted the findings of the supernova groups, and clearly explained the reasons the universe should be flying apart ever faster as astronomers peer deeper and deeper into space. The definitive concept was named the Little Bangs Theory (LBT). In this version, energy is created at the perimeter of the universe expanding outwards at the speed of light. From these energy masses (black holes and quasars) come the ever-expanding numbers of limitless galaxies. In the perspective that everything is connected to everything else in the universe, the LBT led to a FLINE paradigm of planetary origins structured with the Four Laws of Planetary Motion (FL) and internal nucleosynthesis (IN) processes without which evolution (E) is impossible.

But the LBT concept can hold true only if the universe does not fly apart faster than the speed of light; if that speed limit is found to be exceeded, then the necessity of a lambda constant might be likely. But as of now, Einstein appears to be correct in his "biggest blunder" assessment; a lambda constant is not needed here. Although only a small facet of the concept, this exciting "breakthrough of the year" does corroborate the predictions and explanations of the LBT. Rather than searching for lambda or for new physics principles, perhaps a close examination of the 1980 concept would be more productive.

To: William C. Mitchell, Institute for Advanced Cosmological Studies.   3-5-99

Thanks for the copy of your interesting paper, *The Recycling Universe*. I enjoyed reading it, especially the quotes of some great minds of past and recent times (Ref. Jeans, Harwit, Rees, Silk, Davies, etc.). All of the statements fit precisely into the Little Bangs Theory (LBT) of 1980. I could imagine that each of them was speaking of a specific facet of the LBT. And your comments on galaxies fit equally well. In my mind, galaxies consisting of fiery stars, dust-clouds and gases are products of explosions of much larger masses that often place them in bubble-shape formation like points on a spherical perimeter. Only then do the simultaneous processes of accretion and expulsion of matter begin their long cycles.

Your comments on the recycling of heavy elements ring true; however, the question of the original sources of these elements is of paramount importance. We both know that they did not come from the alleged Big Bang (BB). In the perspective of the LBT, atoms are continuously created under variable conditions of extremely high temperatures and pressures within fiery masses of energy. Black holes and quasars, created at the spherical perimeter of our Universe expanding at the speed of light, later eject (usually via explosions) the fiery masses of atom-producing stars buried in dust-clouds. The increasing velocity of the expanding system has been interpreted as being in need of a lambda factor (another fudge factor) to create it *when, in the reality of the LBT, none is needed*. In today's *Atlanta Journal*, Hawking has added his powerful influence to the lambda factor that Einstein correctly discarded as his greatest blunder. Quoting from the article: "*Science* magazine…called the [illusion of an] accelerating universe finding the 'Breakthrough of the Year' for 1998." A sad commentary on science!

One question sure to be asked will concern the true significance of the red shift. Arp has put forth some good arguments against the current interpretation. One comment: time will tell. I agree with the Davies and Gribbin quote "that solitons would appear to us as new types of subnuclear particles…" Further, I consider plasma as an intermediate phase between these particles and our more familiar atomic matter. But I often wonder if dark matter is a figment of the imagination, or simply another fudge factor.

Oxford professor, Dennis Sciama, raises doubt about the BB with his statement: "why the observed spectrum [of MBR] should be that of a black body over a wide range of wavelengths is totally obscure." This crucial fact about MBR is a powerful argument against the BB, and even a more powerful one in favor of the LBT. My book, *The Spacing of Planets*, explains these wavelengths of MBR as emanating from the nucleosynthesis processes within all the fiery spheres (including active planet/moon cores) throughout the Universe. In spite of the claims for it, such processes would not have been possible within the realm of the BB hypothesis, and thus, the MBR cannot be a leftover from a highly improbable explosion.

Bill, to set the record straight, Hoyle, Bondi and Gold named their late 1980s concept the Mini-bangs theory (rather than my Little Bangs) of creation of matter in space. This was done a few years after my LBT manuscript (1980) was copyrighted. Helge's (Ref. 4) *Cosmology and Controversy* (1996), regarding the creation of matter out of the energy of vacuum, was published 16 years after the LBT, which is strongly based on that concept. The first aspect of my paradigm was self-published in 1975 as *Fuels: A New Theory* some five years before Gold published his encroaching *Deep-Earth Gas Hypothesis* (*Scientific American*, 1980). This says something about the intolerance we still endure daily, while others publish, unencumbered by censorship (or plagiarism). John Chappell, Jr. has some strong points in his arguments thereof. For the sake of historical accuracy, won't you consider adding my writings to your credits list?

Of course, your arguments against the BB leave no room for rebuttal. You are doing an excellent job there. My paper on *The FLINE Paradigm: Definitive Insights Into the Origins of Solar Systems and the Orbital Spacing and Evolution of Planets* will follow soon after its final tune-up.

To: PHYSICS TODAY, American Center for Physics, College Park, MD.   05-24-99.

In publishing the two opposing views, *A Different Approach to Cosmology* (PHYSICS TODAY, April 1999) by Burbidge *et al.* and the reply by A. Albrecht, you have done a great service by encouraging open debate in the sciences pertaining to origins and evolution of our universe. While both articles present interesting viewpoints, the different approach offers the more convincing arguments. The most surprising and pleasant aspect, however, is that these frank views were brought clearly into open debate in the same magazine issue — a good idea concerning universal origins that too often seem biased in favor of the BB model.

While the quasi-steady-state universe (QSS) supplies the different and more logical approach, both articles raise questions that appear unanswerable by either of the two models. These questions warrant consideration of a compromising third model capable of supplying these answers. The Little Bangs Theory (LBT), initiated in 1979, offers valid explanations of the controversial issues of a cosmological constant (not needed; Einstein was right), the microwave background, its fluctuations and its connection to ongoing nucleosynthesis throughout the universe (not really possible at any time in the BB concept), the creation of energy and its transformation via nucleosynthesis into the matter comprising all galaxies (the dynamic ejection method) along with many other relevant anomalies.

The principle points of the QSS and the LBT are in general agreement except for the status of the universe. The LBT teaches about an ever-expanding universe growing at the speed of light at its perimeter where energy is created, rather than being created within a steady-state universe. These forms of energy eventually are transformed into the matter comprising all galaxies, which are created, and grow, via the ejection manner described in the QSS version.

The discovery of the proposed Fourth Law of Planetary Motion explaining the spacing of planets in solar systems was accomplished in the years of 1980-1995. Together with Kepler's First Three Laws, the Four Laws (FL) and the internal nucleosynthesis (IN) that is responsible for all planetary evolution (E) offer valid evidence of how our Solar System came into its current stage of existence. All findings on Mars (and other planets) corroborate this FLINE concept of planetary origins and evolution in which the FL, IN and E are chronological, inseparable and ongoing realities of dynamic solar systems. They clearly reveal the necessity of ejection processes in the formation of binary and multiple planets systems while arguing strongly against the condensation concept. Much corroborative evidence in the QSS and the LBT bear powerful witness to this conclusion. All relevant discoveries of the space probes and of Earth continue to interlock precisely into this revolutionary LBT/FLINE concept (initiated in 1973).

The primary hope of this letter is to keep the debate on origins out in the open until we reach the point of absolute proof of how everything came into being. At this writing, the evidence continues to mount against the BB and in favor of the LBT and much of the powerful evidence presented in the article by Burbidge *et al.* Mainstream cosmologists cannot, and must not yet, conclude that we understand the origins and evolution of galaxies, solar systems, planets, etc., via the BB and condensation concepts.

New Principles of Origins and Evolution
Revolutionary Paradigms of Beauty, Power and Precision

To: Glenn Strait, Gordon Rehberg, William Mitchell, File.   07-29-99

The Heart of the Matter (*Science* 2 July 1999 pp55-56), a book review by Roger Blandford, contains information relative to the basic principles of the Little Bangs Theory (LBT) (1973-1980) which state that every galaxy is spawned from the two poles of a central black hole of energy previously created at the spherical perimeter of the Universe during its speed-of-light expansion. The ejecta includes all the forms of energy and matter comprising each galaxy. From such explosive actions come the huge masses of white energy we call stars.

Subsequent interactions of these speeding masses result in many formations of solar systems, usually of a binary nature. In each case, the system must comply with Kepler's Three Laws of Planetary Motion. In the much rarer case of a multiple planets system like ours, two energy masses meet the precise prerequisites of relative masses, relative speeds and precise distance apart to effect a breakup of the smaller, speedier mass into smaller masses as they space themselves in full accord with the Golden Ratio in orbits around the larger mass. *Such systems must comply with all Four Laws of Planetary Motion.* Whether all systems must comply in some way with all Four Laws remains to be determined.

With this background, we can understand why Blandford's review of *Active Galactic Nuclei: From the Central Black Hole to the Galactic Environment* (Princeton University Press 1999), authored by Julian H. Krolik, is an exciting confirmation of the LBT. Other than the confirmation that "...normal galaxies, including our own, contain massive black holes, which occupy about 10^{-30} of their host galaxy's volume," Krolik confirms that quasars, "by converting the gravitational binding of energy of gas just before it is swallowed by a black hole, are quasi-stellar in appearance to the optical astronomer" — as the LBT first stated in an early unpublished manuscript submitted to the Library of Congress in 1980, later published in book form in 1986, entitled *New Concepts of Origins: With White Fire Laden*. Krolik's two confirmations add powerful support to the LBT, yet he appears to cling to the ludicrous Big Bang hypothesis.

Herewith is a copy of the review for your perusal and file. (Note: The second page contained only one-half paragraph, not essential). Any feedback in the form of questions and comments would be appreciated.

To: Glenn Strait, Editor, *The World & I*, Washington, DC 20002   08-31-99

Some time ago, you inquired about my terminology of the composition of the energy cores in universal spheres, specifically that of Earth's core. As stated then, whether we can say that such cores consist of energy particles or plasma or a mixture thereof still remains debatable, and scientists have only to look to these cores to understand the mysteries of how atoms are created within these spheres under severe conditions of high temperatures and pressures. By reversing Nature's processes, scientists are now very close to finding the answers to creation as defined in the LBT/FLINE model.

An exciting article, *Making the Stuff of The Big Bang* (*Science* 20 August 1999, p 1194), by David Voss, unwittingly sheds some light on active core compositions. Voss writes, "If all goes well later this year, physicists [at the Brookhaven National Laboratory]...will create miniature copies of the big bang by smashing together the bare nuclei of atoms traveling at nearly the speed of light. Reaching temperatures a billion times hotter than the surface of the Sun, the protons and the neutrons will melt into their bizarre building blocks: quarks and gluons that hold them together. Out of this inferno will come an exotic form of matter called quark-gluon plasma, primordial stuff that may have been the genesis of all the normal matter we see around us. And most profoundly, the very vacuum, what we think of as empty space, will be ripped apart, revealing its underlying fabric."

The quark-gluon plasma (QGP) concept (with possibly even smaller particles) is what I have always visualized — but until now did not know how to express clearly — as the true composition of the energy cores of all active, self-sustaining spheres of the Universe. The researchers are attempting, albeit unwittingly, to duplicate this core composition. By reversing Nature's processes for creating atomic matter, they will be recreating the QGP from which the atoms were, and are, made within all active spheres, thereby striking at the very heart of ongoing creation. But to move on to the next step in understanding planetary anomalies, scientists must recognize that the QGP processes will go on forever; they did not and could not have happened in the alleged Big Bang. They are indeed the "primordial stuff that may have been [is] the genesis of all the matter we see around us." But rather than from a BB, the origin of the QGP can be best explained via the Little Bangs Theory (LBT) in which the "primordial stuff" will be forever created from "the very vacuum, what we think of as empty space," at the spherical perimeter of the Universe that is forever expanding at the speed of light.

When the QGP-to-atoms concept of creation of atomic matter — the inseparable nucleosynthesis/evolution relationship of the FLINE model — inevitably is recognized and accepted as Nature's basic principle of ongoing creation, the solutions to all planetary anomalies are sure to follow in rapid sequence. The work described by Voss brings scientists to the very forefront of the LBT/FLINE paradigm. After more than two decades of frustration, the exciting breakthrough cannot be too far away!

To: Marilyn Vos Savant, c/o Parade Publications, New York, NY. 01-11-00

Your response to Mike Berman (*Parade Magazine*, Jan 9, 2000) concerning current scientific theories was a pleasure to read. As a scientist (retired), I agree whole-heartedly with the questions raised and the comments made about scientific theories; thanks for sharing these insights.

You "received mail from furious scientists who proceeded to cite every known argument in favor of the Big Bang [BB] theory except the one to which [you] specifically referred: the argument in favor of the entire cosmos once being smaller than a polka dot. That is why I chose it, and its absence from those letters supports my point. ...I also think we must be careful not to teach theories as fact. It slows scientific progress immeasurably." Comment: Furious responses often stem from uncertain consciences.

Not all scientists believe in the BB; many of us are fully aware of its highly speculative structure that stems from evidence based on erroneous assumptions; e.g., a small polka dot beginning. Even one false assumption can lead to many false conclusions. And that is precisely the awkward position in which advocates of the BB have placed science. If the data do not fit an anticipated conclusion, advocates often force-fit it into the concept, softened by "might", "maybe" or "could be" speculation. One example: In recent times, the acceleration of the Universe has been attributed to a repelling force, lambda: a fudge factor to support the erroneous current belief; this, after Einstein called his expansion force the biggest blunder of his life. And the idea that all elements were created in the BB is pure speculation that has no basis in fact; a more logical and provable creation of elements is provided later in this letter.

The BB prevails in spite of its readily refutable evidence and the powerful substantiated evidence against it (ref. *The Cult of the Big Bang* by William C. Mitchell). It does so in the face of a more logical concept that's supported by powerful substantiated evidence, requiring neither a lambda factor to explain the acceleration of the Universe nor a polka dot beginning. These two fallacies pose no problem in the Little Bangs (LB) theory and the subsequent ongoing processes of nucleosynthesis within every active sphere of the Universe. This internal nucleosynthesis (an idea first hinted at by Descartes in 1644) is the driving force behind the evolution of all universal spheres: the two are inseparable — one cannot exist without the other. Many substantiated facts firmly establish these ongoing relationships.

Kepler's Three Laws of Planetary Motion and the new Fourth Law explaining the geometric spacing of planets combine their powerful weight against the Planetary Accretion concept, which, in turn, undermines the BB. Together, these Four Laws (FL) and the internal nucleosynthesis (IN) that drives all evolution (E) form the irrefutable FLINE model of the origins of solar systems and the evolution of planets.

Advocates of the BB should heed your warning that "we must be careful not to teach theories as fact. It slows scientific progress immeasurably." And tragically. As long as the BB and the Accretion hypotheses prevail, a true understanding of universal origins and evolution remains highly improbable.

To: Program Chair, 2001 Meeting, AAAS, Washington, DC 20005   06-29-00
Re: Origins of Solar Systems and the Evolution of Planets.

Science today finds itself in a situation that should be, and must be, resolved before any truly significant progress can be realized in understanding planetary origins and evolution. The tendency toward teaching theories as fact must be carefully avoided while simultaneously keeping open minds to new ideas. As history teaches, when facts displace speculation, new ideas inevitably displace old ones. Your help in bringing these new ideas before the AAAS membership is essential.

I respectfully request that you seriously consider permitting the introduction and open discussion of my work on either one or both new concepts into the main program at the next AAAS Annual Meeting in San Francisco. One pertains to the origin of the universe; the other to the origins of solar systems and the subsequent evolution of planets. The two concepts are intertwined; both remain within the realm of provable facts and sound logic without the need for speculation. Both, in conjunction with current beliefs, were painstakingly researched, developed and carefully evaluated during the past 27 years.

The findings have been summarized in the two pages entitled *Origin of the Universe?* and *Origins of Solar Systems and the Evolution of Planets*. The listed claims of both the FLINE model and the Little Bangs model are backed by substantive evidence; both present strong arguments favoring them over current beliefs. Additionally, my paper, *On the Spacing of Planets: The Proposed Fourth Law of Planetary Motion*, has been in peer review at the Royal Astronomical Society, London, for over 2 1/2 years, apparently without a flaw to justify its rejection. If accepted, of course, it would cause scientists to rethink current beliefs about the origins and evolution of all planets in all solar systems.

Since completion of the proposed Fourth Law (1980-1995) as the final link in the FLINE model, I have been able to reach a number of people who have taken the time and interest to learn and accept its factual message, and consequently have become enthusiastic advocates. But the greater scientific community, who stand to benefit most, remain unaware of its great potential simply because its new ideas have not yet been permitted into the system other than via ineffective poster sessions. These new ideas offer valid and definitive scientific alternatives to the Big Bang and Accretion concepts.

Please examine these findings carefully for any fatal flaw or for any question about the scientific validity of these concepts. If you feel they warrant open discussion and would make an interesting topic at the meeting, please put them on the main program as an invited paper in the planetary sciences. If doubts or questions arise, the feedback from you would be deeply appreciated and helpful.

To: Michael S. Strauss, Program Director, 2001 AAAS Meeting, Washington, DC. 9-4-00.

For the past quarter century, I have carefully researched the origins of solar systems and the evolution of planets from the perspective of prevailing beliefs versus newer concepts. The result is a non-speculative, factual theory. As each facet of the evolving concept became incontrovertible, it was presented (usually in poster sessions} at various annual meetings of the AAAS, AGU, GSA, ACS, GAS and other organizations. The final link that ties everything together and brings the definitive FLINE concept full circle is the solution (1980-1995) to the enigmatic Fourth Law of Planetary Motion clearly revealing how the planets attained their orbital spacing around the Sun. This solution first eluded Kepler in 1595, and until 1995 it had remained the unknown key to understanding the dynamic origins of solar systems. In the sense that it is destined to change scientific beliefs about the way solar systems came into existence, this revolutionary discovery is equivalent to the Copernican idea of a Sun-centered Solar System.

My efforts to bring this important discovery to the attention of the greater scientific community via mainstream scientific programs other than ineffective poster sessions have been thwarted for the past five years by letters that failed to cite a valid reason for the rejections. To circumvent the anticipated rejections, I published my findings in book form to preserve them for posterity. Herewith are two pieces of literature that help convey the significance of these quarter-century findings.

In allowing more logical and accurate interpretations of discoveries in the planetary sciences, the FLINE and LB models have the potential of tremendous benefits to scientists. Poster sessions have proven ineffective, so I see no reason to put these models through another one. For these reasons I responded promptly to your published call for suggestions for the 2001 AAAS Meeting.

With all due respect, I urge you and the program committee members to become familiar with the three fundamental and incontrovertible principles of the FLINE model, so that when its time inevitably arrives, you will permit a symposia presentation to the scientific community by a number of scientists who already understand and accept this revolutionary concept. Now age 77 and with cancer, I would like to see this come to pass in the near future.

From: Brig. Gen. Gordon C. Carson III (Ret), Livingston, Tx.   12-26-00

I am the son of Gordon and Eleanor Carson whom you met thru Opal Falligant. When I was home for Christmas, my parents showed me a copy of your book, The Spacing Of Planets. During my stay in Savannah, I read it several times and find it to be the most brilliant and fascinating work I have seen in many years.

The Big Bang Theory has never been acceptable to me. In my opinion, the universe has always been here and will always be here as a perpetual motion machine. The age of the oldest galaxies may simply represent the normal lifespan of such bodies. And as you point out, millions of new galaxies are continuously being created at the perimeter of our expanding universe.

Another point in which I agree is the solar origin of the planets and moons. The iron core theory has not been convincing, and your belief of the internal nuclear furnace makes perfect sense, owing to the composition of the Sun.

There are many other points which I find intriguing and would enjoy discussing with you, but the reason for my letter is to ask how I might buy four copies of your publication… one for myself, and one each for three friends. Please send purchasing information to my Georgia address below.

To: Dr. William C. Mitchell, Chairman, Institute for Advanced Cosmological Studies, Carson City, NV 89702   06-25-00

Just in case you missed the recent article, *The Universe in a Sphere,* in Parade Magazine, I'm sending a copy. After you have read it, I'm sure you'll understand why I felt compelled to comment on the brand-new Hayden Planetarium that is its centerpiece. Exhibits depict the development of the Universe from the "Big Bang" (BB) on. Its Rose Center is planning extensive nationwide educational programs via various news media, thereby making scientific findings, data, interpretations, etc., available to classrooms and communities across the land.

It is unsettling to learn that so much effort and expense is being put forth to advance the BB as a valid scientific concept, especially when there is far more substantiated evidence against the BB and far too much speculative interpretations used in support of it. What a terrible waste! And how embarrassing it will be to science in the not too distant future when astronomers learn the true nature and origins of comets and asteroids.

Revved-Up Universe: Astronomers check out an expansive finding (*Science News*, Feb 12, 2000), states: "Just two short years ago, two teams of astronomers presented the first evidence that we live in a runaway universe, driven to expand at a faster and faster rate." You may recall my letter of February 20, 1998 in which I explained that this accelerating expansion was first predicted and explained in my 1980 version of the Little Bangs Theory (LBT) in which the Universe is expanding at its perimeter at the speed of light. This naturally explains why the Universe appears to astronomers to expand at a faster and faster rate as they look farther and farther out.

"That finding is in direct conflict with the simplest version of the BB," the article continues. "According to that theory, the universe has expanded ever since its explosive birth, but gravity has gradually slowed the expansion. Even if the universe grows forever [a principle of the LBT], the theory predicts that it should do so at a steadily decreasing rate. Recent observations of exploded stars, however, suggest that the universe's rate of expansion is, in fact, increasing [another principle of the LBT]. Over the past year [and as predicted by the LBT], new data appear to [do] corroborate those findings."

However, the reason given by astronomers for the apparent increase in the rate of expansion requires a fudge factor: the cosmological constant that Einstein discarded in later years by calling it his greatest blunder. There is no need for the cosmological constant in the LBT; the apparent increasing rate of expansion is a natural consequence, a basic principle, of the LBT.

The reasons I'm writing this, other than to be sure you are aware of the new Hayden Planetarium and its educational programs, is to get this load off my shoulders and into the written record for future generations; also, to record that not all scientists believe in the unsubstantiated BB. To me, it does seem wrong to push the BB into people's minds without more definitive evidence to support it. But of course, such evidence will always be elusive.

P.S. - Astronomers appear to be very familiar with the LBT; they seem headed in that direction.

To: Brig. Gen. Gordon C. Carson III, Conyers, GA 30013   12-29-00

Thank you for the tremendous compliment expressed about my latest book, *The Spacing of Planets,* and for the four-book order in your letter of December 26. By the time you receive this response, you should have received the books mailed to your Georgia mailing address, given by Eleanor by phone the evening prior to receiving your letter.

You may already know that the inspiration for the research efforts that ultimately led to a series of books on origins and evolution stemmed from a conversation with your father in the early 1950s speculating about the origin of our Moon. This occurred at the Falligants' (Claude & Opal) home on Talahi Island between card games. After almost 20 years of cogitating and raising a family had passed, the pieces began interlocking precisely into place. In-depth research soon proved, beyond doubt, the abiogenic origins and intimate relationship of the hydrocarbon fuels, first self-published in 1975 as *Fuels: A New Theory*. This revolutionary concept proved to be the crucial key to unlocking the mysteries of how all planetary systems came into being and to the factual understanding of planetary anomalies. In turn, this led to the enigmatic solution to the Fourth Law of Planetary Motion that first eluded Kepler in 1595.

The new Fourth Law (1980-1995) was the final link that brought the FLINE model full circle. Sadly, as it turns out, my real work had only begun. During the past two decades of the evolving model, each new facet was presented in various science meetings to national and international groups of scientists. Since its completion in 1995, it still accumulates rejections by peer reviewers schooled in current beliefs based on the erroneous Big Bang (BB) hypothesis.

To put the BB situation in perspective, I'm taking the liberty of sending to you a copy of *The Cult of the Big Bang: Was There a Bang?* by William C. Mitchell. In my opinion, this is the most authoritative and factual book ever written on the subject. Its contents mesh perfectly with my Little Bangs/FLINE model. However, since revolutionary ideas are upsetting to BB believers who exercise control over scientific content (press releases, etc.), we both experience the same rejection problems. So, we join with many other scientists in attending the Natural Philosophy Alliance meetings, an international group who meet one or two times a year to discuss, with open minds, all new ideas that challenge prevailing beliefs.

A copy of page 10 of the AAAS meeting February 15-20, 2001, is enclosed. The four subjects checkmarked will be based on the BB; other concepts are not permitted on their main program — only in ineffective poster sessions. Thus, progress in the planetary sciences will remain stymied.

The Spacing of Planets now can be found on the Internet at www.iuniverse.com, by typing in author's name. The updated version that embraces relevant findings of the past five years is due online in late January or February. You will find three of its pages enclosed.

With your permission, I would like to use, as a promotional blurb, the statement from your letter: "I read the book several times and find it to be the most brilliant and fascinating work I have seen in many years."

To: Glenn Strait, Science Editor, The World & I, Washington, D.C. 03-07-01

Re: ON THE SPACING OF PLANETS: The New Fourth Law of Planetary Motion.

Because exciting breakthroughs within the realm of natural laws, along with the rapid accumulation of corroborative evidence, assure the scientific validity of this manuscript, I'm taking the liberty of sending it for your consideration for publication.

If too long as is, its separate, but interlocked topics, with editing, can be conveniently published in any combination, thereof, to meet any specification of length for an interesting and historical article. For example, the *Abstract, Introduction and The Geometric Solution to the Fourth Law of Planetary Motion* could be used as one article. The addition of *Extrasolar Systems: How and Why They Differ From Our Solar System* would add more interest and only two additional paragraphs.

The section headed *The Five Stages of Planetary Evolution* appears capable of standing alone, but could be enhanced by a connection with the manner in which the planetary nebulae attained their orbital spacing around our Sun before beginning their evolution through the five stages. The same reasoning could be applied to *Corroborative Evidence for Nucleosynthesis Within Planets and Moons* and to *The FLINE Paradigm of Planetary Origins and Evolution*.

These are only suggestions — probably superfluous — which you, as editor, can accomplish better than I could. At any rate, I hope you will give the contents the full consideration they warrant, and will feel free to edit as you wish. As more people read about these breakthroughs, the list of excited advocates grows (see blurbs) along with my confidence that the concept contains no fundamental flaw — especially since no one has managed a valid rebuttal during the past 28 years of its evolution. But there is one caveat: my paper on *The Spacing of Planets* has been in peer review for slightly more than three years at the Royal Astronomical Society, which gives me even more confidence in its scientific validity.

Isn't it reasonable to believe that a fundamental flaw, if any, would have been discovered and exposed during the many years of the FLINE model's development and discussions with knowledgeable scientists? Its potential benefits to science are immense. It is time to move on to the next paradigm shift as recognized by Thomas Kuhn — one that finally offers provable solutions to the anomalies of solar systems and planets. Imagine the tremendous controversy in which your magazine could reap the benefits.

Self-explanatory letters concerning its status accompany the manuscript. You have the option for first rights to magazine publication; however, I will retain these and all other rights if, for any reason, you decide not to publish it soon.

To: Dr. William C. Mitchell, Institute for Advanced Cosmological Studies, Carson City, NV.   March 16, 2001

Re: Space observatory shows black holes once dominated. (*Atlanta Journal*, 03-14-01)

This article discusses the findings of the Chandra X-ray telescope. To quote: "Huge black holes once dominated the universe — sucking in gas, dust and stars, erupting with surges of X-rays that have journeyed since for billions of years across the heavens. That's a picture of the early universe captured by the orbiting Chandra X-ray telescope in a study that focused on a small section of the sky for days-long exposure to capture faint X-rays streaming from more than 12 billion light-years away."

"The Chandra data show us that giant black holes were much more active in the past than at present," Riccardo Giacconi, a John Hopkins University astronomer, said Tuesday at a news conference." Added Bruce Margon, a professor at the University of Washington, Seattle: "If you look at the sky with X-ray eyes, you see almost nothing but black holes." Proponents of the Big Bang (BB) believe that the study confirms theories by showing that the early universe teemed with active black holes, spewing X-rays across the heavens, and that Chandra is looking at the X-ray universe as it existed up to 12 billion years ago. But they do not say exactly which theories are confirmed.

However, when the evidence is examined more closely, it presents very powerful arguments favoring the Little Bangs (LB) over the BB theory. The evidence is precisely as predicted and explained in my book, *The Spacing of Planets: The Solution to a 400-YearMystery* (pp 111-126). The fact of ever greater numbers of black holes seen upon approaching the universe's rapidly expanding perimeter is a basic principle of the LB. They are the ultimate source from which all galaxies are spawned. This fact explains both the ever greater outward velocity of galaxies as astronomers look ever farther into space and the close relationship of greater distances to younger galaxies. All other relevant anomalies become vulnerable to easy solutions via this revolutionary LB/FLINE model — a concept solidly underpinned with the Four Laws of Planetary Motion. Its supporting evidence is incontrovertible.

Thus, the conclusion that the Chandra data show us that giant black holes were much more active in the past than at present is not justified; one can conclude only that their ever-increasing numbers verify the rapid growth of the universe via the creation of the densest form of energy (black holes) at its ever-expanding perimeter. When viewed in this light, the data bring astronomers ever closer to the LB/FLINE model of ongoing origins and evolution while pushing them ever further from the BB.

Please feel free to use any of this information in your next book.

To: Sen. Newt Gingrich, 1301 K Street, NW, Washington, DC 20005   05-11-01

As a concerned R&D scientist (retired), I would like to offer some belated comments on your fine article, *An Opportunities-Based Science Budget* (*Science* 17 Nov 2000), and hope that perhaps they are not too late to do some good. I am in total agreement with the five major increases in federal funding for scientific research that you feel should be seriously considered in the science budget.

Your suggestion number three (3) is of special interest: "Money needs to be available for highly innovative, 'out-of-the-box' science. Peer review is ultimately a culturally conservative and risk-avoidance model. Each institution's director should have a small amount of discretionary money, possibly 3% to 5% of their budget, to spend on outliers. The history of plate tectonics should remind all of us that accepted wisdom can be wrong." Proper use of discretionary money could alleviate some very costly situations in science.

Most scientists realize that throughout a history beginning with the Copernican idea of a heliocentric solar system, the accepted wisdom of the time has almost always been proven wrong. And today's beliefs already have been shown to be no exception to this great truth. Any suggested change in the direction of scientific thought, as with Copernicus, still meets with persistent and unfair peer resistance, almost always through peer review by advocates of prevailing beliefs. No new idea that challenges current beliefs, regardless of its scientific validity, can get past advocates in control of peer review and grant money. My personal experience during the past quarter-century certifies this point.

History clearly reveals that dramatic breakthroughs always occur in the mind of one person rather than via consensus of many minds. Science finds itself in the situation of being, once again, at the crossroads of decision. Revolutionary evidence of new paradigms as irrefutable as that of Copernicus cannot get past peer review. Your wise suggestion of discretionary money to spend on outliers seems to be the only hope for revolutionary ideas that can point us in the right direction of scientific thought in the planetary sciences. The remaining choice is to continue to spend billions of dollars teaching and trying to prove current speculative and erroneous hypotheses. You have touched upon the most crucial issue now facing the planetary sciences, and perhaps other sciences. Enclosed are some pages of self-explanatory literature that illustrate the issue at hand.

Thank you for the timely article; I hope this information is not too late to be of significance to the cause.

P.S. - NASA space science chief, Ed Weiler, is right to put the freeze on the Pluto mission (*Science* 17 Nov 2000) — at best, an ego trip. Scientists can learn nothing of real significance that's not already known about this planet; the $1.4 billion price tag would be close to a total waste. If acquiring highly significant knowledge is the goal, the money would be spent far more profitably on two other missions: to Europa and to the next nearby comet. The latter mission will reveal the most shocking and revolutionary results; perhaps this could be the reason such a mission has been delayed as long as possible.

To: Donald Kennedy, Editor-in-Chief, Science, AAAS, Washington, DC 20005.   5-28-01

Re: The Five Fundamental Principles of Origins and Evolution of Planetary Systems.

The literature herewith includes a brief synopsis of the definitive FLINE model embracing the five fundamental principles of origins and evolution of planetary systems. The fifth and final principle — the enigmatic solution to the new Fourth Law of Planetary Motion that first eluded Kepler in 1595 — inexorably links with the first four principles of the revolutionary FLINE paradigm to bring it full circle to an irrefutable conclusion. This geometric solution clearly reveals, beyond any reasonable doubt, how the planets attained their orbital spacing around our Sun. The new model provides a solid foundation for a full and complete understanding of all anomalies of solar systems.

Recent examples of the FLINE model's potential for elucidation of new data are found in the seven articles on *Comet C/1999 S4 (LINEAR)* (*Science* 18 May 2001). The strange behavior of the disappearing comet leaves many unanswerable questions, solely because the interpretations of all data are based on the prevailing belief (posing as fact) in the speculative "snowball" nature of comets. Interpretations of such data in the perspective of the realistic "fireball" nature of comets never encounters the problem of unanswerable questions. The FLINE model leaves no room to doubt the true nature and sources of comets, or why they disappear, and why the total mass measured following the breakup of Comet C/LINEAR is about 100 times less than the estimated total mass prior to its breakup.

Any of the intermeshed facets of the FLINE model can stand alone on its own merits. Each of them tells a factual story of crucial significance to the scientific and world communities; e.g., *The Energy Fuels (Non-Fossil) Theory, The New Fourth Law of Planetary Motion, The Fiery Nature and Sources of Comets and Asteroids, The Dynamic Origins of Solar Systems, The Five-Stage Evolution of Planets,* and the subject paper: *The Five Fundamental Principles of Origins and Evolution of Planetary Systems.* In conjunction with its many interlocked facets, the fifth principle predictably — as with the Copernican idea — will force dramatic changes in beliefs about our planetary systems in which all five of these inseparable principles, each dependent on the other four, already are playing equally significant roles.

A page of old and new blurbs adds assurance of public interest in, as well as the scientific validity of, these new perspectives, all carefully researched during the past three decades to eliminate myths and speculation. A manuscript of specified length on any facet of the new model is available upon request.

Thank you for your time and consideration to publish this short version of *The Five Fundamental Principles of Origins and Evolution of Planetary Systems*. I fully comprehend the situation it entails, and urge you to pursue these crucial findings for the sake of pure science, the AAAS creed, and posterity.

The Genesis Mission to the Sun: Scheduled for July 30, 2001 - Sept 2004.

The news article, *Spacecraft to Catch a Piece of the Sun. Bring it Home*, by Marcia Dunn, AP (*Atlanta-Journal-Constitution*, July 15, 2001), stated: "A robotic explorer named Genesis is about to embark on an unprecedented journey to gather and bring back bits of the Sun. NASA hopes the specks of solar wind — equivalent to perhaps 10 grains of salt, if added up — will help to explain what it was like in the beginning. The very beginning. when the planets in the solar system were forming."

"...As the solar wind streams past the spacecraft, microscopic traces of chemical elements will become embedded in materials on collector panels. These castaway elements are the same material as the original solar nebula, the disk of gas and dust from which all the planets and all other solar system objects formed 4 1/2 billions years ago. The outer layers of the sun, continually streaming into space as solar wind, provide the most feasible way to access this fossil record." The spacecraft is on schedule for a July 30 launch, and should return to Earth in September, 2004 with the small samples.

But there are caveats. Stephen Hawking, a prominent physicist, recently made a stunning admission: "Cosmology's hypotheses are conjectural when pressed to the limit." True; they are purely speculative. As a matter of fact, definitive evidence against these prevailing concepts has become overwhelmingly persuasive; e.g.. the five principles of origins and evolution revealing that planets did not, and do not, accrete from a disk of gas and dust. As divulged by the five principles, they began as moderate-size masses of nuclear energy that had broken from a separate, but smaller, faster-moving, mass that passed close to the larger Sun, and had been forced into geometrically-spaced orbits around it. Most are still undergoing the processes of evolution via the transformation of their internal energy into matter, all in full accord with Einstein's ubiquitous $E = mc^2$ and with Nature's natural laws. Some smaller ones have exhausted their energy to become inactive (dead) planets (e.g., Mercury, Mars, Pluto).

The Four Laws of Planetary Motion assure that, like the planets, our Sun began as a nuclear energy fireball, larger than now, but much as we see it today: a huge fiery mass of ejecta from a much larger energy mass (black hole or quasar). They clearly confirm that the dynamic interactions between the two energy masses gave birth to our Solar System. Here, no conjectural cosmology is needed; the plethora of supportive evidence appears incontrovertible.

Should the Genesis mission's objectives be viewed in this new perspective, rather than in the current one, the futility of the mission becomes obvious. The captured pieces of the Sun will be virgin material created within its massive interior and ejected into space, as is true with all fiery stars. This material will be the elemental products of internal nucleosynthesis, which occurs in all active spheres of the Universe — and it will be basically the same (selectively qualitative, but not quantitatively equal) as all other ejecta material from other active spheres. Its significance will have no relevancy to the alleged Big Bang origin of our Universe or of the Accretion of planets — both appear to be erroneous conjectures.

But perhaps the results will prove beneficial in advancing scientific knowledge about the Sun's internal nucleosynthesis, and thereby, increase our knowledge about the first stage of evolution of all planets: the energy stage.

A TIME FOR CHANGE

A quarter century of painstaking research into the origins of solar systems and the evolution of planets has failed to uncover sufficient substantiated evidence in support of the Big Bang (BB). However, the research did uncover sufficient substantiated evidence against the BB and in powerful support of an ongoing Little Bangs (LB) concept of creation. In these contrasting concepts, the LB remains steadfast in its capability to explain the stream of enigmas encountered in the BB concept.

The LB, supported by the new Fourth Law of Planetary Motion and Kepler's First Three Laws, presents powerful arguments against the concepts of planetary accretion, cores of iron, rock or silicate in active planets, primordial asteroids, snowball comets, BB nucleosynthesis, a cosmological constant, finite energy and fossil fuels. Each of these antiquated theories is structured with speculation and misinterpretations, with no basis of substantiated facts. Each is fundamentally flawed, a house of cards destined to crumble in light of the LB and its FLINE model's three intimate and inviolate principles: the Four Laws of Planetary Motion (FL) and the ongoing universal nucleosynthesis (IN) that continually drives all evolution (E).

In contrast to the BB, the LB and the revolutionary FLINE model offer viable alternatives that factually circumvent fundamental flaws and give assurance that current antiquated concepts can not long endure. For example, the recent discovery that our Universe is expanding at an ever faster rate as astronomers look ever farther into space is very powerful evidence against the BB, which predicted just the opposite: the ever slowing rate of expansion at ever greater distances. Simultaneously, this crucial discovery — an inherent principle of the LB concept — is powerful substantiated evidence for the ongoing LB theory that accurately predicted and explains why the expansion rate increases with distance. No cosmological constant is necessary.

The recent exciting discoveries of two groups of young planet-sized objects unattached to any mother stars, along with the 50 known extrasolar systems, were predicted and explained by the LB as early as 1980. Perhaps recognition of the LB/FLINE concept will come when scientists learn the true nature and sources of asteroids and comets in the first decade of this century.

But the path will not be easy. Obviously, the strong entrenchment of the prevailing Accretion Disk theory (the dust aggregation/planetesimals/accretion hypothesis) in the scientific literature presents one of history's most formidable barriers to change in direction of scientific thought. However, history does teach that unsatisfactory concepts are replaced eventually by more satisfactory ones.

ABOUT THE AUTHOR

Alexander Scarborough grew up on a farm in Macon, Georgia during the Great Depression of the 1930s. In close touch with nature, he developed a deep curiosity about how everything came into being. When his seventh grade textbook taught the fossil fuel concept, Scarborough could not comprehend how plants and animals could possibly account for the world's deposits of gas, oil, and coal. Additionally, his observation of lumps of coal revealed the presence of gas and oil inside them. But what impressed him most at the time were the Copernican idea of a Sun-centered Solar System and the clockwork precision of Kepler's Three Laws of Planetary Motion. His goal in life was fixed.

Studies at Lanier High School and later at the University of Georgia, Athens, where he attended on an SAT scholarship, were concentrated in chemistry, physics, and math. After graduation in 1944, he served two years in the Army, with a stint in the Po Valley campaign in Italy. His education was furthered at Georgia Tech via a degree in chemical engineering, 1947-1949, followed later in life by self-taught geology and astronomy sciences. Through membership in a number of science organizations (e.g., AAAS, AGU, ACS, AICHE, etc.) throughout his career in Industrial Research & Development, and during his decade of retirement, he has kept in close touch with the frontiers of the sciences pertaining to the universal inanimate matter/energy relationship.

A resident of LaGrange, Georgia, Scarborough raised a family of three children. After his retirement on January 1, 1990, he concentrated his efforts on gaining additional knowledge, both factual and speculative, about planetary origins and evolution. His discovery of the revolutionary Fourth Law of Planetary Motion (1980-1995) on the spacing of planets around our Sun, has proven to be the missing link that finally ties everything together, while further exposing a number of fallacies and myths of science; e.g., the antiquated fossil fuels hypothesis that had sparked his lifelong quest for understanding how everything came into being. Simultaneously, the discovery added powerful support for the seemingly indisputable energy fuels and FLINE concepts as valid alternatives to prevailing beliefs.

New Principles of Origins and Evolution is the ninth work in his Energy Series, initiated in 1975 with *Fuels: A New Theory*. Each is an expanded version of all previous editions; each is enhanced with recent discoveries of the time. During the quarter century of evolvement into a revolutionary LB/FLINE model that now flows with a continuity of beauty, power, and precision, this new concept has compiled an impeccable record of accurate predictions and definitive explanations of the startling discoveries of science. Scarborough believes that his lifelong work finally has brought the Copernican Revolution (the understanding of our Solar System) full circle.

CPSIA information can be obtained
at www.ICGtesting.com
Printed in the USA
LVHW061449301021
701976LV00014B/405